"绿都北京"研究系列丛书
Green Beijing Research Series

北京北中轴区域绿色网络规划与设计研究

Planning and Design of Green Network Renewal in the Areas of North Axis, Beijing

北京林业大学园林学院
王向荣 刘志成 林箐 等　编著

中国建筑工业出版社
CHINA ARCHITECTURE & BUILDING PRESS

图书在版编目(CIP)数据

北京北中轴区域绿色网络规划与设计研究/王向荣等编著. ——北京：中国建筑工业出版社，2019.6
("绿都北京"研究系列丛书)
ISBN 978-7-112-23523-0

Ⅰ.①北… Ⅱ.①王… Ⅲ.①绿化规划-研究-北京 Ⅳ.①TU985.21

中国版本图书馆CIP数据核字（2019）第055347号

责任编辑：杜　洁　李玲洁
责任校对：姜小莲

"绿都北京"研究系列丛书
北京北中轴区域绿色网络规划与设计研究
北京林业大学园林学院
王向荣　刘志成　林箐　等　编著

*
中国建筑工业出版社出版、发行(北京海淀三里河路9号)
各地新华书店、建筑书店经销
天津图文方嘉印刷有限公司印刷
*
开本：787×1092毫米　1/16　印张：11　字数：348千字
2019年6月第一版　2019年6月第一次印刷
定价：99.00元
ISBN 978-7-112-23523-0
　　　（33810）

版权所有　翻印必究
如有印装质量问题，可寄本社退换
(邮政编码 100037)

编 委 会

主　　编　王向荣　刘志成　林　箐
副 主 编　李　倞　钱　云　王沛永
　　　　　　段　威　尹　豪　张云路

本研究获北京市共建项目专项资助"城乡生态环境北京实验室"、北京林业大学美丽中国人居生态环境研究院和国家自然科学基金项目（31600577）共同支持

前 言

美国著名风景园林师西蒙兹（John Simonds 1913-2005）出版过一本文集Lessons（中文版书名为《启迪》），书中有一篇文章记录了西蒙兹1939年到北平考察的经历。

在北京，西蒙兹拜访了一位祖上曾经参与规划了元大都的李姓建筑师，李先生非常赞赏他能到北京考察风景规划，并向西蒙兹简单地介绍了元大都的规划思想。"在这片有良好水源的平原上，将建设一个伟大的城市——人们在这里可以与上天、自然以及同伴们和谐共处"。"蓄水池以自然湖泊的面貌贯穿整个都城，挖出的土用来堆成湖边的小山，湖边和山上种植了从全国各地收集来的树木和花灌木"。"关于公园事宜和开放空间，可汗命令不能有孤立的公园。更准确地说，整个大都城将被规划成一个巨大而美丽的花园式公园，期间散布宫殿、庙宇、公共建筑、民居和市场，全都有机地结合在一起"。"从文献中我了解到北平被一些来此旅游的人称为世界上最美丽的城市，我不知道这是否正确。如果真是这样，那么这种美丽不是偶然形成的，而是从最大的布局构思到最小的细节——都是通过这样的方法规划而成的。"

北京的确如西蒙兹在文章中提到的李姓建筑师所说，是一座伟大的城市，也是一个巨大而美丽的花园式都城。

北京有着优越的地理条件，城市的西、北和东北被群山环绕，东南是平原。市域内有5条河流，其中的永定河在历史上不断改道，在这片土地上形成广阔的冲积扇平原，留下了几十条故道，这些故道随后演变为许多大大小小的湖泊，有些故道转为地下水流，在某些地方溢出地面，形成泉水。

北京又有着3000年的建城史。李先生提到的元大都已将人工的建造与自然环境完美地叠加融合在一起，到了清朝时期，北京城人工与自然的融合更加紧密完善。城市西北郊建造了三山五园园林群，西山和玉泉山的汇水和众多泉流汇纳在一起，形成这些园林中的湖泊，水又通过高粱河引入城市，串联起城中的一系列湖泊。许多宫苑、坛庙、王府临水而建，水岸也是城市重要的开放空间。城中水系再通过运河向东接通大运河。由此，北京城市内外的自然景观成为一个连贯完整的体系，这一自然系统承担着调节雨洪、城市供水、漕运、灌溉、提供公共空间、观光游览、塑造城市风貌等复合的功能。城市居住的基本单元——四合院平铺在棋盘格结构的城市中，但每一个四合院的院子里都有别样的风景，每个院子都种有大树，如果从空中鸟瞰，北京城完全掩映在绿色的海洋之中。

然而，随着人口的增加和城市建设的发展，北京的环境在迅速地变化着，古老的护城河已部分消失，一起消失的还有城市中的不少湖泊和池塘。特别是快速城市化以来，北京的变化更为剧烈。老城中低矮的四合院被高楼大厦取代，步行尺度的胡同变成了宽阔的道路。老城之外，城市建设不断向周边蔓延，侵占着田野、树林和湿地，城市内外完整的自然系统被阻断，积极的公共空间不断消失，而交通设施的无限扩张，又使得城市被快速路不断地切割，城市渐渐失去了人性化的尺度、也渐渐失去了固有的个性与特色。

面对自然系统的断裂和公共空间的破碎与缺失等城市问题，作为风景园林、城市规划和建筑学的教育和研究者，我们看到了通过维护好北京现有的自然环境和公园绿地，利用北京的河道、废弃的铁路和城市中的开放土地，改造城市快速交通环路，建设一条条绿色的廊道，并形成城市中一个完整的绿色生态网络，从而再塑北京完整的自然系统和公共空间体系的巨大机会。

这一绿色的生态网络可以重新构筑贯穿城市内外的连续自然系统，使得城市的人工建造与自然环境有机地融合在一起；这个网络可以将由于建造各种基础设施而被隔离分割的城市重新连接并缝合起来，形成城市的公共空间体系；这个网络可以承载更加丰富多彩的都市生活，成为慢行系统、游览、休憩和运动的载体，也成为人们认知城市、体验城市的场所；这个网络还可以带来周边地区更多的商业机会，促进周围社区的活力；这个网络更是城市中重要的绿色基础设施，承担着雨洪管理、气候调节、生态廊道、生物栖息场构建、生物多样性保护的关键作用。

这套丛书收录的是我们对北京绿廊和生态网络构建的研究和设想。当然，畅想总是容易的，而实施却面临着巨大的困难和不确定性，但是我们看到，世界上任何伟大的城市之所以能够建成，就是从畅想开始的，如同元大都的建设一样。

在《启迪》中那篇谈到北京的文章最后，西蒙兹总结到："要想规划一个伟大的城市，首先要学习规划园林，两者的原理是一样的"。

我们的研究实质上就是以规划园林的方式来改良城市，希望我们的这些研究成果也能对北京未来的建设和发展有所启迪。

2018 年 1 月

Forewords

The famous US landscape architect John Simonds once published a corpus named 'Lessons', and one of the articles in this book records the experience of Simonds's investigation to Beijing in 1939.

Simonds visited an architect surnamed Li whose ancestors once took part in planning of the Great Capital of Yuan in Beijing, and the architect admired that Simonds came to Beijing to study landscape planning. He also briefly introduced the planning thoughts of the Great Capital of Yuan to Simonds's group. According to architect Li, here on this well-watered plain, was to be built a great city in which man would find himself in harmony with God, with nature, and his fellow man. Throughout the capital were to be located reservoirs in the form of lakes and lagoons, the soil formed their excavations to be shaped into enfolding hills, planted with trees and flowering shrubs collected from the farthest reaches of his dominion. As for the matter of parks and open spaces, architect Li said the Khan decreed that no separate parks were to be set aside. Rather, the whole of Ta-Tu would be planned as one great inter-related garden-park, with palaces, temples, public buildings, homes and market places beautifully interspersed. He also added that he was led to believe that Peking (now present day Beijing) was regarded by some who have travelled here to be the most beautiful city of the world, which he could not know to be true. If so, it would be no happenstance, for from the broadest concept to the least detail --- it was planned that way.

Just like what architect Li mentioned, Beijing is indeed a great city, also a grand gorgeous garden capital.

With superior geographical condition, Beijing city is surrounded by mountains in the west, north and northeast direction, and the southeast of the city is plain. There are 5 rivers in the city. Among them, the Yongding River has constant change of course in history, thus formed the vast alluvial fan plain here and has left dozens of old river courses. These old water courses then evolved into lakes with different scales, some even transformed into underground water streams and overflowed to the ground to form springs.

At the same time, Beijing has a history of 3000 years of city construction. As architect Li said, the Great Capital of Yuan has integrated the artificial construction and the natural environment perfectly. And when it came to Qing Dynasty, the integration of labor and nature is even more perfect in Beijing city. People built the 'Three Hills and Five Gardens' in the northwest of the city, so that the catchments of the West Mountain and Yuquan Mountain could join numerous springs together, and formed the lakes in these royal gardens. Then, water was introduced into the city through the Sorghum River, and thus a series of lakes inside the city are connected. Plenty of palatial gardens, temples and mansions of monarch were built in the waterfront, which makes the water bank an important open space for the whole city. The river system in the city heads for the east and connects to the Grand Canal, which makes the nature environment inside and outside the city into a coherent and complete system, which takes the charge of compound functions including the regulation of rain flood, city water supply, water transport, irrigation, providing public space, sightseeing function and shaping the cityscape. As the basic unit of urban living, Siheyuans are paved in the city with chessboard structure. Uniformed as they are in appearance, we can still see unique landscape and stories in each different courtyard. There are big trees thriving in each courtyard, as if they were telling the history of each family. If we have a bird's eye view from the air, Beijing will be completely covered in the green ocean.

However, with the population increase and the urban construction development, the environment of Beijing has been changing rapidly. The ancient moat has partly disappeared, together with many lakes and ponds in the city. Beijing has changed even more fast and violent since the rapid urbanization. Low Siheyuans have been replaced by skyscrapers, and Hutongs of walking scale also became broad roads for vehicles. Apart from the Old City, the urban construction in Beijing has been spreading to the surrounding area, invading the fields, forests and wetlands. As a result, the holistic natural system both inside and outside the city is blocked, active public space is disappearing, and the unlimited expansion of transportation facilities make the city constantly cut by express ways. We cannot deny that the city has gradually lost its humanized scale, and it has also gradually lost its inherent personality and characteristics.

In face of the city fracture problems of natural systems and the broken public space, as landscape architects, urban planners, architecture educators and researchers, we see huge opportunities to maintain the existing natural environment and garden greenways, use the river courses, disused railways and open lands in Beijing to reform the city fast traffic roads and construct several green corridors in order to form a complete green ecological network in the city, and remold integrated natural system and public space system in Beijing.

This green ecological network can reconstruct the continuous natural system running throughout the city, so that the artificial construction of the city can be organically integrated with the natural environment. The network can connect and stitch the city divided by all kinds of infrastructure, and form a public space system in Beijing. What's more, the network can carry more colorful urban life styles and become the supporter of slow travel system, sightseeing, recreation and sports, and it will turn into a place for people to cognize and experience the city. It can also bring more business opportunities in the surrounding areas to promote the vitality of the communities in the neighborhood. Above all, the network is a significant ecological infrastructure in the city, which plays key roles in rain and flood management, climate regulation, ecological corridors, biological habitat construction and biodiversity conservation, etc.

This collection includes our researches and thinking of greenways and the construction of ecological corridor network in Beijing. It is without doubts that imagination is always easy, while implementation is always faced with great difficulties and uncertainties. But we can see that any great city in the world was finally built up based on the imaginations in the beginning, just like the construction of the Great Capital of Yuan.

In the article about Beijing from 'Lessons', Simonds concluded that: If you want to plan a great city, you need to learn to plan gardens first, for the principles of both are the same.

Essentially, our research is to explore a way to improve a city in the way of planning gardens, and we do hope that our research results may enlighten the future construction and development in Beijing.

Wang Xiangrong
January, 2018

目录 / contents

01 规划设计
PLANNING & DESIGN

"U"计划　　　　　　　　　　　　　　　　　　　　13
基于更新理念的城市开放空间设计
U plan
Urban Open Space Design Based on the Conception of 'Urban Renewal'

蓝绿交织　　　　　　　　　　　　　　　　　　　33
基于适宜性分析的城市生态网络构建
Intertwining Blue and Green
Construction of Urban Ecological Network Based on Suitability Analysis

"绿地+"　　　　　　　　　　　　　　　　　　　53
基于复合型绿地整合的北中轴绿网更新
Green space+
The Renewal of Beijing Central Green Network Based on Green Space Integration

共享绿色的实现　　　　　　　　　　　　　　　　75
渗绿小空间体系构建
Sharing Green
Green Space Penetration System Construction

生态绿网，活脉融城　　　　　　　　　　　　　　97
北京市北中轴生态网络更新构建研究
Green Lines as City Veins
The Establishment of Ecological Networks around the North Central Axis of Beijing

向"芯"生长 119
北京北中轴绿网更新
Centripetal Growth
Renewal of Beijing Central Green Network

生活的赞礼 139
北京北中轴绿网更新
The Praise of Life
The Green Network Renewal of the North Central Axis of Beijing

"帧"渗透 153
北京市中轴绿网更新研究
"Frame" Infiltration
Research of Green Network Renewal in Beijing Central Axis Area

02 研究团队
RESEARCH TEAM

核心研究团队 171
Core Researchers

特邀专家 173
Invited Experts

研究生团队 175
Postgraduates

01 规划设计
PLANNING & DESIGN

进入新世纪，中国的城市发展也开始进入新的阶段。在过去的三十余年，由于中国城市化率严重不足，一直以来更加注重城市化的数量增长，快速城市化使中国的城市化率迅速提升。这种城市化模式为中国的经济发展带来了巨大的推动力量，但是随着城市史无前例地快速扩张，也产生了中国城市病的严重问题。目前，中国的城市化发展已经进入新的分水岭，城市化进程开始更多地转向质量提升。这点在诸如北京、上海、广州等中国特大城市中表现得尤其明显。量质共生应当是未来中国城市化发展的主旋律。

高品质城市的一个重要特征就是提升城市的公共服务质量。不同的学科都在探索中国新型城市化的质量提升模式。从风景园林的视角来看，建设高品质的城市绿色公共空间系统是获得高质量城市服务的重要内容，对城市品质提升效益非常显著。过去，大家往往更加关注扩张型城市（城市新区）的公共空间建设。如今，已建成区的绿色公共空间系统建设将被提上日程。此类公共空间建设具有很强的特殊性和挑战性，此时城市已经完成基本建设，空间制约非常明显。风景园林需要利用城市更新的机遇，挖掘未充分利用的城市空间潜力，从而开展设计。此类空间的利用需要结合场地现状的精细化分析手段和创造性的设计介入模式。同时，由于用地比较复杂，需要与城市规划、建筑、交通、水利等多学科开展更加密切的合作。

本课程题目聚焦于北京的北中轴线及周边城市区域。该区域是北京城市发展中具有特殊历史意义和重要价值的区块。北二环路及护城河是明清京城古城遗址的一部分；青年湖、柳荫公园等是北京早期公园建设的代表性成果；中华民族园、奥体中心（公园）、奥林匹克中心区等是改革开放后大型活动引领建设的标志性成就。本次规划设计工作在北京市"城市功能疏解"、"绿道建设"、"2008奥运遗产再利用"和"2022冬奥会筹办"的大背景下，试图通过精细化的区域调研、发掘绿色更新潜力空间，提出合理、可实施的空间更新策略，构建一个从北京老城区向北延伸至奥林匹克中轴线的北京中央城区绿色公共空间网络。

Since the 21st century, the development of Chinese cities entered a new phase as the response to the long-term low urbanization rate in the past. Consequently, the key themes of such a process focus only on the speed of the physical growth, which may generate great push power of local economy but also potential "urban disease" in many aspects. This means the coming "turning point" of China's urbanization, and the quality improvement of urban spaces will be the main task in future. In some larger metropolises like Beijing, Shanghai, and Guangzhou, the historic changes are emerging right now.

There have been tremendous findings in pursuing appropriate ways to create higher-quality urban spaces. From the perspective of landscape architecture, the establishment of green open space system should be significant measurements to improve the quality of urban spaces. This includes not only the construction of greenways in new urban areas, but also effective betterments of the existing urban areas. Certainly, the planning and design of the renewal process must be more unique and challengeable, and require deeper investigation on the existing and potential green areas and needs of local community, usually with more detailed analytical tools and innovative design. Also, this should be a complex process under close cooperation by urban planners, architects, civil engineers and hydraulician and so on.

This project focuses on the area around the north central-axis of Beijing city. This is a very special area which just reflects the urban development trajectory of Beijing. The North 2nd ring road and the moat are parts of the historic town; the Qingnianhu Park and Liuyin Park are two typical cases of early city parks in the Socialist planned-economy era; while Chinese Ethnic Culture Park, the Olympic Green and sports parks are the symbols of recent urban construction since the era of the reform and opening-up policy. Currently, under the idea of "better reuse the Olympic heritages" and the new master plan of Beijing Municipal City, the renewal process aims to regenerate a green open space system within the area in very cautious ways. The work expects to create green connections across the historic town, Socialist modernist city and the Olympic central area along the north central-axis of the capital city.

The site is located on the north of Beijing Second Ring Road and the west of the central axis of the city, which is the junction area of Xicheng, Haidian, and Chaoyang District. The connection of the block to the wedge-shaped green structure in the urban area of Beijing has positive significance for the continuation of the ecological corridor, and also plays an important role in enhancing the experience of the city.

Under the background of urban 'double repair'——the urban repair and ecological restoration, the renewal of green network in the old city is a key step in the reconstruction of the old city, which is of great significance to the improvement of the old city space quality and the living environment. The renewal of the green network in the old urban areas is mostly based on the preservation of the existing green space, which usually means the minimum intervention alone with the best achievements. Given the situation that many Chinese cities is carrying out the totally reconstruction of the old city, it is worth to rethinking how to update the old city network scientifically on the basis of maintaining the old urban features.

The construction of the site began from the 1960s, and now is dominated by residential areas. Most of the buildings were traditional courtyards. Among those courtyards, transportation facilities cover a large proportion of land which has potential renewal value. The site is attached to the North Second Ring Road and there are two community parks, and a water system is running through the site as well. However, the green space is distributed fragmentedly and structurally weak, and the amount of greenery on the west side is obviously insufficient. At the same time, because this area was constructed about fifty years ago, although the internal road network system and public transportation of the site are relatively complete, the road is narrow, the parking facilities are insufficient, and the roads are usually occupied for other purposes, which results in poor road slowness and comfort. In addition, historical spot and cultural heritages such as Taiping Lake, Zhuanhe River, Deshengmen Gate Tower, etc., will become important nodes for transformation and design.

In general, the location advantage of the site is obvious, and it is strongly associated with the adjacent area, but the internal land use is chaotic, the traffic is complex, and the green quantity is insufficient, so it needs to be optimized and integrated. The plan aims to increase the total amount of green space, establish a complete green network to connect important nodes, and then establish a comfortable slow traffic system. It will also create distinctive open spaces for different site conditions.

Therefore, the following design strategies are proposed: Firstly, art galleries will be constructed along the central axis; secondly, historical and cultural recreation areas will be built along Zhuanhe River and North Moat; then, the Axis Art Gallery, the second ring road greenway, and the Yuan Ruins Park will be linked to form a U-shaped greenway to create a comfortable slow traffic system. In this way, the internal micro-circulation of the site will be open, and the potential space will be fully esplored to stimulate the site's vitality. In the end, a "one axis, one belt, three centers" planning structure will be formed.

Specifically, we divide this site into 'three centers' as axis art district, the riverside cultural district and the neighborhood leisure district along the North central axis and Zhuan River, which are called 'the Axis' and 'the Belt'. A perfect green network is formed through the pertinent transformation and design. The linear system forms the skeleton of the green network to meet the demand of walking and cycling respectively. And the transportation station will be combined to form a perfect system of public service facilities, including the tourist service station, recreation facilities, parking facilities as well as health facilities and so on, which form a U-shaped green way. Finally, the Slow-traffic System of the old city area will be built. Dot and block system mainly include public spaces such as commercial center, art center, cultural center and green land of parks. We put forward corresponding transformation modes for different nodes, so as to create multifunctional living areas and to meet the residents' demand of daily activities, and to stimulate the vitality of this area.

Finally, the ecological corridor will be continued by completing the structure of Beijing green network through the green network skeleton, and the internal microcirculation will be formed by extending the green network to inner city, then public space will be created to meet the demands of residents' leisure activities by creating historic city landmarks to improve the living quality of residents, and finally the sustainable development of the city will be realized.

"U" 计划
基于更新理念的城市开放空间设计
U plan
Urban Open Space Design Based on the Conception of 'Urban Renewal'

许少聪、杨宇翀、罗雨薇、逯羽欣、刘心梦、耿菲
Xu Shaocong / Yang Yuchong / Luo Yuwei / Ti Yuxin / Liu Xinmeng / Geng Fei

中轴线在北京城市规划中具有"城市地标"的特殊意义，其周边用地的发展影响着北京城市特色的体现。

场地位于北京市二环路以北、中轴线以西以及西城区、海淀区和朝阳区的交界地带，在生态、休闲等方面具有重要作用。场地内既有人定湖公园、北滨河公园等开放活动空间，也有太平湖、德胜门等重要历史文化节点，但由于交通割裂、用地局限等原因，场地内部景观状况参差不齐，现状绿地早已无法满足居民日益增长的休闲文化需求。

此次规划力求建设一条城市生活绿色纽带，在绿道骨架的支撑下，延续绿色网络，打通城市"微循环"，进一步打造绿色核心，激发场地活力，创造一个兼具多样性、开放性、生态性和实用性的城市空间，实现区域的整体可持续发展。

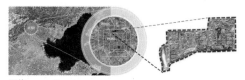

区位
Location

规划地块地位于北京市北二环以北，中轴线的西部，属于西城区、海淀区、朝阳区三个城区的交界地带。规划地块大规模建设开始于20世纪60年代，多是单位大院，具有封闭性、复杂性和复合性的特色，大部分地区属于旧城区，建设空间有限，发展限制因素较多。

地块位于北京城区楔形绿地结构末端，对于生态廊道的延续具有积极意义，同时对提升游赏网络体系也具有重要作用。在建设之初，场地中水系主要包括太平湖、转河，后来由于地铁二号线太平湖车辆段的建设，如今太平湖已被填埋。

目前，用地性质以居住用地为主，由于北京北站和太平湖车辆段以及德胜门公交总站的存在，交通设施用地占地面积大，具有较大的潜在更新价值；地块内有两个社区公园和北二环绿道，并有贯穿东西的水系，但绿地分布破碎且分布不均，导致地块西侧绿地总量明显不足；场地公共交通设施较完善，但修建年代早，道路狭窄，停车场设施不足，停车占道现象严重，慢行舒适度和穿行体验较差。

总体来说，区位优势明显，区块各类资源丰富，潜在更新价值较大，但用地相对混乱、交通复杂、绿量不足，绿网需进行整合。

根据前期研究，规划设计在满足慢行系统舒适度的基础上，建立绿色网络，通过规划、设计和管理的线性绿色空间廊道，串联场地内零散的潜在绿色空间，形成一个连续的空间网络。借助地块区位优势，明确场地基本定位，提出四条主要设计策略：沿中轴打造艺术长廊；沿转河、北护城河建设历史文化休闲廊道；连通中轴艺术长廊、环二环绿道、元大都城垣遗址公园形成"U"形绿道，构建慢行系统；打通居住区内部微循环，充分挖掘潜力空间，激发场地活力。

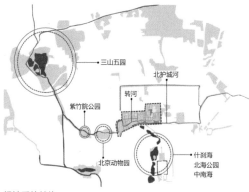

绿地系统结构
Green Space System Structure

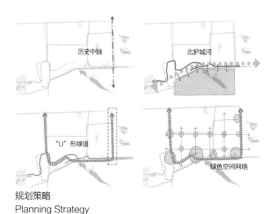

规划策略
Planning Strategy

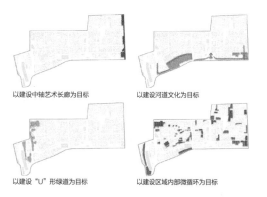

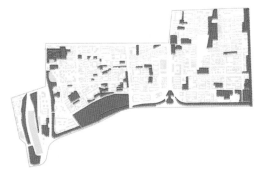

潜力地块选择
Potential Block Selection

规划设计 PLANNING & DESIGN

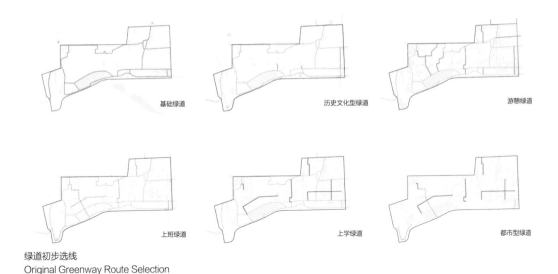

绿道初步选线
Original Greenway Route Selection

绿道初步选线叠加
Overlaid Original Greenway Route Selection

潜力地块与绿道选线叠加
Overlaid Potential Block and Greenway Route Selection

规划以目标为导向，分别针对四条策略选取四类潜力地块。①以建设中轴艺术长廊为目标，作为北京北中轴起始的一部分，打造艺术文化中心；②以构建北二环护城河河道文化区域为目标，挖掘北护城河和城墙的历史文化价值，打造城墙文化长廊；③以建设"U"形绿道为目标，将中轴艺术长廊区域、北护城河区域元大都城垣遗址公园连成一体，形成完整的区域绿道体系；④以建设区域内部微循环为目标，部分拆除社区围墙，形成社区之间的微循环体系。最终再比较分析四类潜力地块，加以校正，保证潜力地块选择的合理性。

绿道的选择采取以目标导向为主、问题导向为辅的方法，分为两个部分。①根据场地重要因子（场地内外的绿地水系以及已选好的潜力地块）构建绿网骨架线路；②在绿网骨架的基础上，针对不同的需求，选取包括以服务居民为主的都市型绿道线路和以服务游客为主的历史文化型绿道线路，在都市型绿道中，根据居民的不同需求，选取不同的影响因子，进而选出以服务居民休闲为主的游憩绿道、以服务居民上下班为主的上班通勤绿道和以服务中小学生为主的上学通勤绿道。在绿网骨架选线的基础上，叠加都市型绿道和历史文化型绿道，进行比较分析与优化，最终获得绿道选线，形成整个地块的绿道体系。

将潜力地块与绿道选线结果进行叠加、校正，舍弃与绿道联系性较差、更新价值较低的潜力地块，去掉与绿道联系性较强、但拆改难度较大的地块，最终划定更新范围。

规划设计 PLANNING & DESIGN

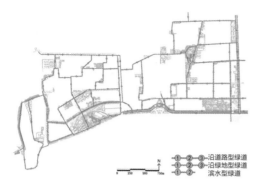

绿道类型
Greenway Type

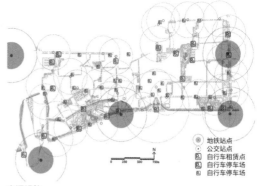

交通设施
Traffic Facilities

绿道分级
Greenway Grading

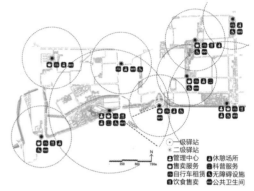

驿站 & 服务设施
Post Station & Service Facilities

绿道选线确定后，我们将绿道分为三类：沿道路型绿道、沿绿地型绿道和滨水型绿道。根据绿道设计导则及现状，布置场地内自行车停车和租赁点，设置自行车驿站，规划服务设施，完善功能布局。

我们对绿道提出了分类改造策略。第一种是沿道路型绿道，细分为余地较大、较有余地及余地很小三种类型。对于余地较大的道路，结合建筑前广场，设置活动、休闲空间，扩大原有绿化带，设置侧向停车位，靠近人行道一侧设置自行车道。具体设计如下：设置人行道宽度 >2m，自行车道 3m，铺装采用透水材质；行道树间距 6m，采用冠大荫浓树种；分车带采用耐性强灌木，间隔 6m 种植行道树，广场上设置座椅等休息设施；间隔 24m 设置路灯、12m 设置高杆灯、12m 设置草坪灯，树下设置树灯、30m 设置垃圾桶。行道树间设置自行车停车场，采用嵌草铺装，并注重无障碍设计。对于较有余地的道路，人行道内侧设置休闲座椅，外侧设置自行车道。对于余地较小的道路，可设置垂直绿化及路侧雨水花园。

第二种是沿绿地型绿道，同样依据余地大小分为三种类型。根据不同情况，设置面积大小不一的休闲空间，有余地处设置雨水花园，沿道路设置休憩座椅。

第三种是滨水型绿道，分为沿滨河公园型和利用滨水车行道改造两类。公园内有余地处设置两级道路，距离水面较近处设置亲水步行道，外侧设置自行车和步行混行道；第二类车行道侧设置自行车道，结合滨水阳台扩大步行区域，同时提供观景、休闲空间。

对于绿道与交通干道、河道等的交接点，我们提出四种处理模式。第一种为平交模式，适用于交通繁忙程度较低、道路不宽的区域，必要时可设置交通岛作为缓冲。第二种为过街天桥模式，适用于交通复杂的核心区域，连接周围公园、绿地、轨道交通出入口、重要建筑及其屋顶（形成屋顶绿色花园），成为城市观景阳台，激发区域活力，促进城市更新。第三种为地下通行模式，适用于河道、跨度较大的机动车道。结合休息、活动、灯光设施，激发场地活力。第四种为城市综合体模式，结合地下交通，形成城市三维空间，解决交通、功能、景观等多种需求，成为城市地标。

我们在前期调研的基础上，将规划区域分为七大类，分别为：居住生活类、生态科普类、体育类、休闲游憩类、历史文化类、商业办公类及儿童活动类，对不同类型区域提出模式化的改造策略，打造多功能活力生活区。

中轴及滨河条带为规划重点区域。中轴作为城市主要景观节点联系的走廊，赋予艺术功能。滨河条带以地铁二号线太平湖车辆段的改造项目为契机，打造新的文化商业中心，同时连接环二环绿道，提升沿河一带的场地活力。

最终通过分片区针对性的改造设计，形成一个完善的绿色网络，其中线性系统构成了绿色网络的骨架，并依据绿道性质与级别分别满足步行与骑行要求，结合地块内交通站点布置自行车租赁点与驿站，完善公共服务设施系统。

线性系统与绿地斑块连接形成一个绿地综合体，提升规划地块内绿地质量并增加绿量，发挥生态、娱乐、文化和美学等多种功能，并在城市尺度下连接北京城区楔形绿地结构，延续生态廊道。

边界U形绿道依次通过商业中心、艺术中心、文化中心与公园绿地激发场地活力——U形绿地中心点，接入北京中轴体系与环二环城市绿道体系，形成城市地标。

类型一：平交模式
交通繁忙程度较低、道路宽度不大等区域采用平交形式，必要时可设置交通岛作为缓冲。

类型二：过街天桥模式
在交通复杂的核心区域可架设景观天桥，连接周围公园绿地、轨道交通出入口、重要建筑及其屋顶（形成屋顶绿色花园），成为城市阳台，激发区域活力，促进城市更新。

类型三：地下通行模式
河道、跨度较大的机动车道等可设置地下穿行通道。结合休息、活动、灯光设施，激发场地活力。

类型四：城市综合体模式
重点区域可建设城市综合体，结合地下交通，形成城市三维空间，解决交通、功能、景观等多种需求，成为城市地标。

绿道与交通节点衔接模式
Connection Mode of Greenway and Traffic Node

居民生活类　　生态科普类　　历史文化类　　商业办公类

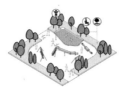

体育类　　休闲游憩类　　儿童活动类

活动类型
Activity Type

规划设计 PLANNING & DESIGN

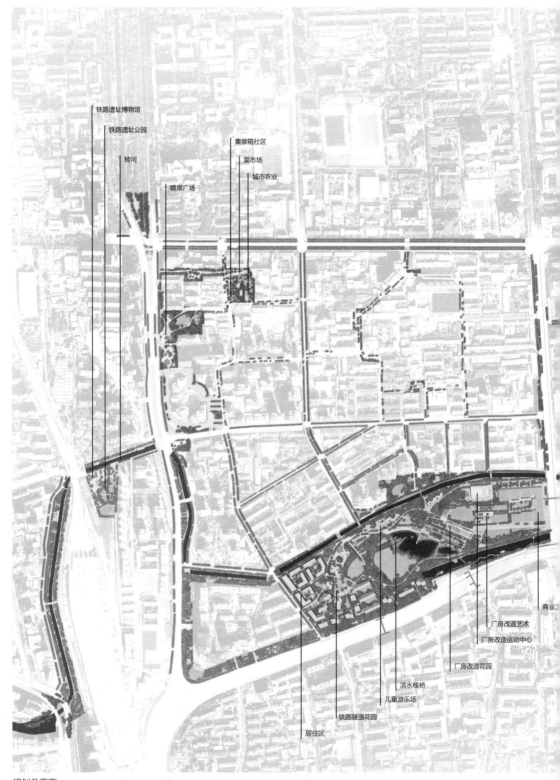

规划总平面
Master Plan

规划设计 PLANNING & DESIGN

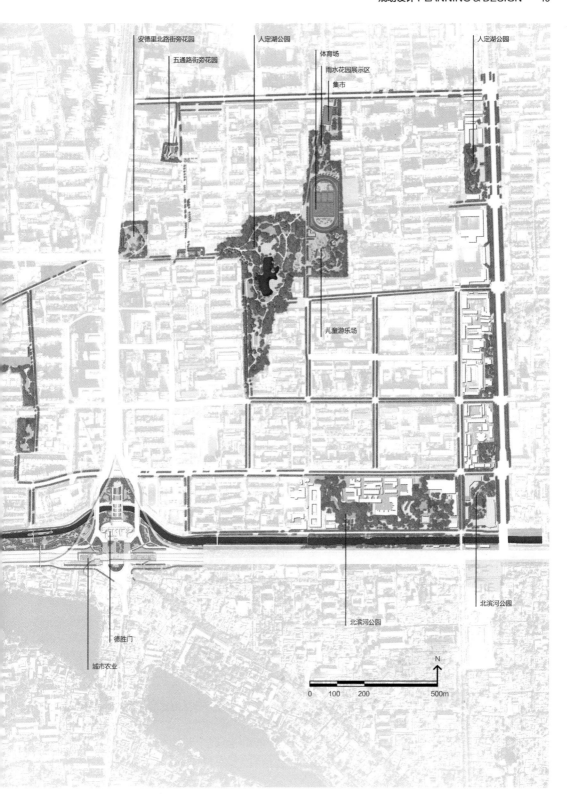

1. 中轴焕活

中轴区域设计平面
Design Plan of Central Axis

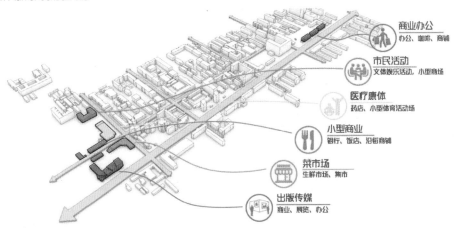

周边区域功能布局
Arrangement of Peripheral Function

中轴部分北段有封闭的军区大院，中段以居住区为主，南接北二环城市绿道，南段功能以菜市场及小商铺为主。

根据总体规划，中轴需要完成强化中轴气氛、完善绿道网络的目标。中轴部分作为呈现北京市城市规划的重要节点，需要具有承上启下、兼具文化复兴的作用，体现城市美感。内部功能以发展文化创意高新产业为主，兼具商业、绿地、办公、居住等功能，打造具有艺术氛围的街道。场地不同区域存在不同的主要问题。

北段军区大院造成空间封闭，阻隔了场地南北商业的连接。中段外部功能不合理，存在大量停业商铺，内部街区面貌混乱，慢行空间不足。南段北滨河公园封闭，与场地开放的定位不符，菜市场与周边商铺距离居住区较远，街区整体相对封闭，活动空间利用率低。同时由于高差的存在，使北二环城市绿道与鼓楼外大街的自行车联系中断。

分析认为场地整体交通不便，功能布局存在不合理，通过对部分建筑进行改造，同时对中轴线空间进行整体设计，在北京中轴线区域自北向南依次设计。自北向南分别为商务办公区、临街休憩区、艺术展览区、社区中心、健身活动区及滨河景观带六大功能区。

北部拆除废弃汽修厂并增加绿地，沿街规划写字楼，增加商业功能，增强南北街区的功能连续性，提升街区承载力，形成街道——商业写字楼——室外活动空间——休憩绿地的功能序列，并美化围墙。

中部将停业店铺拆除或整改，使小型商铺集中在内部居住区，削弱鼓楼外大街旁的商业功能，置换成街旁绿地，营造舒适街道环境，地铁站周边增加自行车服务点。

南部北滨河公园的改造旨在改善现状功能单一、边界封闭的问题，增强边界开放性，增加活动空间，内部引入娱乐休闲功能。同时，重新定义东南角地块功能，内部解决自行车穿行问题，地块西北角整合菜市场功能，使其贴近居住区，方便市民生活，东南角引入出版传媒综合体，整合周边大量出版传媒产业，形成联系更加紧密的传媒活动中心。

最终，场地边界被引入新的混合功能，建筑功能包括商业办公楼、市民活动中心、菜市场、临街小型商业建筑和出版传媒综合体，其中商业办公楼与周边形成统一界面、菜市场进行更新与位置调整、出版传媒综合体进行功能梳理，得到功能分区，并提供屋顶花园进行活动。

为了完善绿道网络，提升慢行系统舒适度，中轴地段由外到内进行了道路断面改造，根据道路通行能力与规划定位要求布置功能。

地块边界为鼓楼外大街、北二环城市绿道，该道路现状通行能力较好，因此改造旨在遵循现状宽度的前提下满足步行、骑行的舒适度，布置休憩空间、增加服务点等，最终形成绿道网络骨架。

地块内部街道则毗邻居住区，现状条件不一，改造方式主要结合现状，重新梳理功能，布置停车空间，合理规划各功能所需道路宽度，保障步行、骑行的通行能力，构建通勤绿道、游憩绿道，完善绿道网络系统。并在断面改造时，增加道路透水铺装与生态草沟，提升绿道网络的生态性。

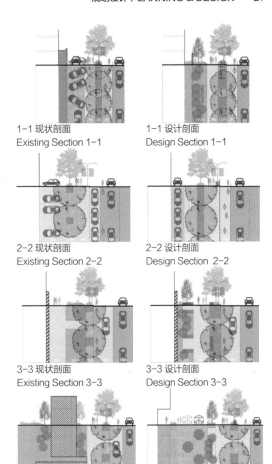

1-1 现状剖面
Existing Section 1-1

1-1 设计剖面
Design Section 1-1

2-2 现状剖面
Existing Section 2-2

2-2 设计剖面
Design Section 2-2

3-3 现状剖面
Existing Section 3-3

3-3 设计剖面
Design Section 3-3

4-4 现状剖面
Existing Section 4-4

4-4 设计剖面
Design Section 4-4

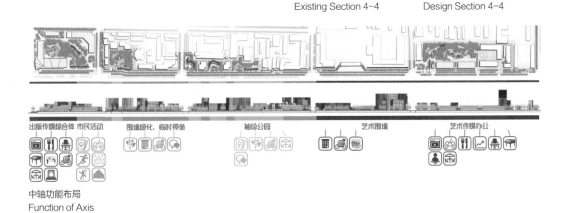

中轴功能布局
Function of Axis

艺术围墙效果
Perspective of Art Walls

袖珍公园效果
Perspective of Pocket Park

出版传媒综合体效果
Perspective of Media HOPSCA

休闲步道效果
Perspective of Recreation Trail

2. 德胜门复兴

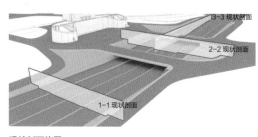

现状剖面位置
Section Location

1-1 现状剖面
Section 1-1

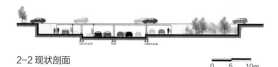

2-2 现状剖面
Section 2-2

德胜门复兴段的规划目标是沿北护城河建设历史文化休闲区。德胜门段交通复杂，以机动车为主的交通规划缺乏对慢行体系的考虑，使得人车混行，慢行体系不完善。虽然德胜门具有深厚的文化底蕴，但经过 500 余年的发展，其所承担的城市功能逐渐丧失，目前作为古钱币博物馆对外开放。

将德胜门作为切入点，从绿道体系梳理和城市功能赋予两方面来探讨未来德胜门地区发展策略。

首先对公交进行梳理。将公交总站沿八达岭高速向北搬迁到北沙滩，该地区靠近地铁站和上清桥交通枢纽，并正在进行开发。再对公交线路进行梳理，整合、废除、新建公交站点，最终场地将有 8 个站点。对二环下沉部分采取盖板策略，通过对主路与辅路标高的分析，得到落柱位置，进一步确定盖板范围。

整个设计分为德胜门城墙记忆段、滨河文化休闲段、德胜门文化广场段。城墙记忆段设计中，在二环上方采取恢复一段城墙的策略，运用现代的手法对城墙元素进行提取、设计。并将城墙赋予一定的城市功能，如咖啡厅、文化展馆、剧场等。城墙还是绿道体系的一部分，通过立体交通的构建，满足步行与自行车基本的穿行功能。同时城墙兼具游览功能，让人体验德胜门的文化底蕴。在滨河文化休闲段中综合考虑城市各级交通，保证地铁、公交与场地的便捷，加强城市与河道的联系性，并结合河道设计一系列有特色的滨水开放空间节点。

最终德胜门地区在城市当中将是整个绿道体系的一部分。在城市功能方面，弥补城市中的公共设施，重新焕发的德胜门地块的活力。

3-3 现状剖面
Section 3-3

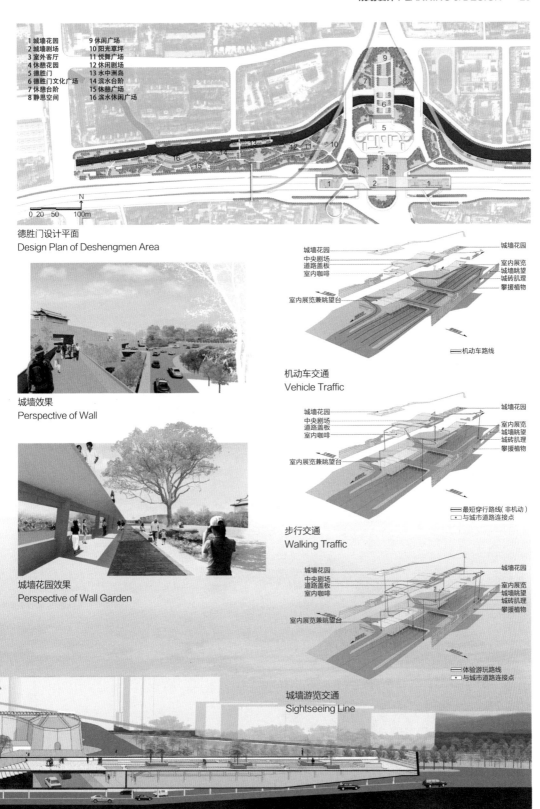

3. 太平湖车辆段激活

在上一阶段的规划中，将太平湖车辆段定位为充满活力的绿色空间。集商业、文化、办公、居住休闲娱乐等功能于一体，在满足社区居民使用的同时，吸引更大范围的人群，营造城市景观和构筑文化场所，即通过构建一系列的美术馆、媒体中心、展示馆、影院、露天剧场、音乐广场和博物馆等富有地域文化特色的休闲空间，表达丰富的城市文化内涵。

场地存在三点主要问题：硬质铺装面积大，缺少绿地；场地封闭，可达性弱；缺少基础设施，人群活动单一。以上三点现状问题导致场地缺乏生机与活力。

城市工业文明带来以上问题的同时，也为场地保留了独特的气质，成为设计最大的机遇和挑战。根据规划定位的大目标，并结合现状实际条件，制定了四大策略：构建公共绿色空间网络，激发场地生命活力，保留历史工业美感，打开边界与城市连接。

设计分为四个区域：便民生活区、居住区、太平湖公园区、文化商业休闲区。

便民生活区：该区块规划了集商业、办公、居住、公共服务为一体的多功能建筑组团，临街对外的建筑具备商业、办公等功能；对内的建筑有居住、社区服务等功能，社区服务中心内有托儿所、老年活动中心、健身中心等。

居住区：基本保留现状的居民楼，并对其破碎的现状公共空间进行整合，主要以休憩绿地为主，配备一些健身运动等服务设施。

太平湖公园区占据了场地的大部分面积，是区

滨水广场效果
Perspective of Waterfront

铁轨花园效果
Perspective of Track Garden

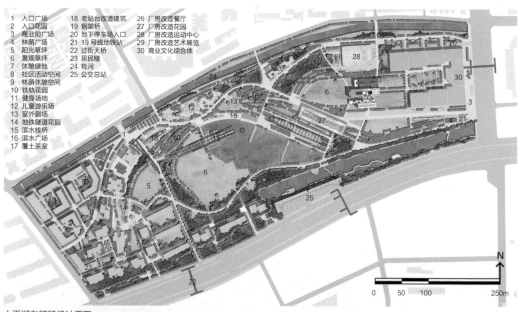

1 入口广场
2 入口花园
3 商业前广场
4 林荫广场
5 阳光草坪
6 景观草坪
7 休憩绿地
8 社区活动空间
9 林荫休憩空间
10 铁轨花园
11 健身场地
12 儿童游乐园
13 室外剧场
14 地铁隧道花园
15 滨水栈桥
16 滨水广场
17 覆土茶室
18 老站台改造建筑
19 钢架桥
20 地下停车场入口
21 19号线地铁站
22 过街天桥
23 居民楼
24 转河
25 公交总站
26 厂房改造餐厅
27 厂房改造花园
28 厂房改造运动中心
29 厂房改造艺术展览
30 商业文化综合体

太平湖车辆段设计平面
Design Plan of Taipinghu Park

域的中心绿核。公园对外开放，可达性强，与其他三个分区联系密切，方便居民进入。与此同时，公园的活动场地多样，功能丰富，景观类型多样，配套服务设施齐全。公园更是保留了铁轨、厂房等要素，并对其进行改造，形成具有特色的景观。

文化商业休闲区：引入一个文化商业综合体，提高土地利用价值，聚集人气，激发场地活力。建筑主要包括购物中心、文化艺术展馆、多媒体中心、图书馆、办公空间等。同时预想未来的 19 号线地铁站将设置在场地的东南角和购物中心内部，可以为场地带来更多活力。建筑南侧为公园入口，有一条艺术廊道，廊道两侧设置有艺术雕塑、景观灯、休闲廊架等景观元素，将与建筑对接形成一条动感活力的艺术步道，将市民引入公园内部。

公共绿地结构
Structure of Public Green Space

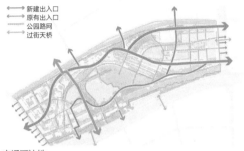

交通可达性
Transportation Accessibility

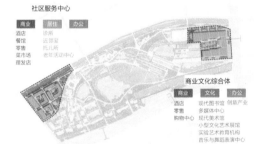

建筑功能布局
Layout of Building Function

社区中心
Community Center

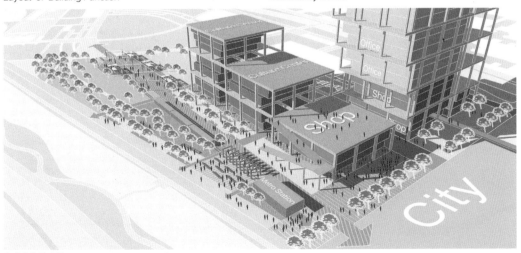

商业文化综合体
Commercial and Cultural Center

4. 转河两岸改造

转河周边地块在上位规划中属于滨河文化区，主要目标是沿转河、北护城河建设历史文化休闲区域。此区域周边交通、水系、绿地资源丰富，经转河向西可连接三山五园绿道，向北可连接海淀十字绿廊、元大都城垣遗址公园等。1905年，京张铁路的修建使得原本的河道向北折行，形成如今的转河。

基于现状分析，提出四项解决策略：慢行串联、活力聚集、场所记忆和生态连接。方案分为滨水生态段、城市休闲段、邻里生活段和铁路遗址公园四个分区，多个主题功能区沿廊道分布。

滨水生态段：偏自然风貌的河道景观，沿河岸设置了人工湿地、垂钓、林荫广场等不同的功能区，同时将原本封闭的地铁十三号线高架桥桥下空间打开，营造出一条具有艺术氛围的长廊。

城市休闲段：以硬质驳岸为主。河岸结构分为上下两层，上层与城市道路齐平，景观以植物造景结合场地为主；下层为一条贯通的滨水廊道。

邻里生活段：主要服务周边居民，在更新时注入了儿童游戏、草坪剧场、健身广场等功能。

铁路遗址公园片区：借助场地废弃铁路的资源优势，保留现状铁轨，并在铁轨上放置一些车厢，用作可移动的临时服务建筑。同时增设铁路遗址博物馆，注入家庭园艺等功能。

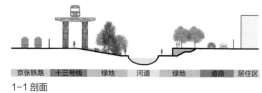

1-1 剖面
Section 1-1

2-2 剖面
Section 2-2

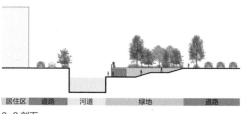

3-3 剖面
Section 3-3

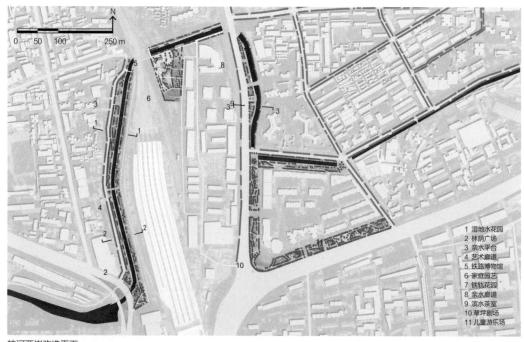

转河两岸改造平面
Design Plan of Zhuanhe Renewal

规划设计 PLANNING & DESIGN   27

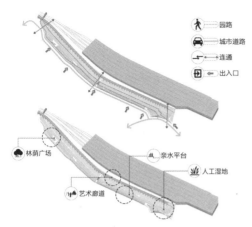

滨水生态段
Analysis of Waterfront Ecological Section

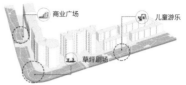

邻里生活段
Analysis of Neighborhood Section

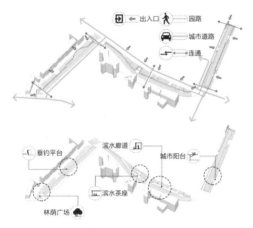

城市休闲段
Analysis of Urban Leisure Section

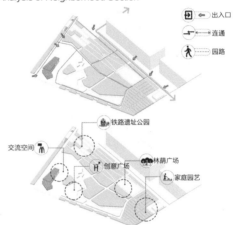

铁路遗址公园
Analysis of Railway Heritage Park Analysis

滨水生态段效果
Perspective of Waterfront Ecological Section

邻里生活段效果
Perspective of Neighborhood Section

城市休闲段效果
Perspective of Urban Leisure Section

铁路遗址公园效果
Perspective of Railway Heritage Park

5. 学院南路社区更新

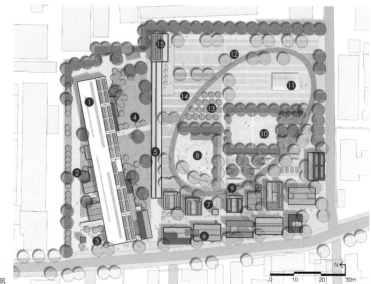

1 菜市场　6 多功能街区　11 温室
2 卸货区　7 健身步道　12 作物区
3 入口广场　8 露天剧场　13 果树区
4 休闲草坪　9 休闲广场　14 花卉区
5 景观廊架　10 门球场　15 服务建筑

社区更新平面
Design Plan of Community Renewal

　　元大都至太平湖片区位于规划区域西北角，属于海淀区，场地在上一阶段规划中功能定位属于邻里休闲区。场地西北临元大都城垣遗址公园，西南临转河，东南临规划的太平湖公园，以封闭居住区和大院为主。商业方面，场地内有一个大型购物中心，其他多为超市、沿街底商及单层商铺。文教历史方面，场地内有中影影城、大学及小西天牌楼。交通方面，场地内部无地铁站，有少量公交站点。

　　该区域现存主要问题：场地内绿量不足；大院内活动丰富，但应用率极低，公共空间激发的活动种类单一；封闭小区及大院阻隔了交通，与周边绿地连接性不强，缺少公共交通，慢行系统不连贯；机动车、自行车占道现象严重；私搭乱建严重。

　　设计承接规划定位，对城市公园进行连接。挖掘场地潜力，增加绿色节点，如社区公园、开放空间、街角绿地等，并通过改造街道、规划慢行系统，达成构建社区绿色微循环体系的目标。

　　对于社区慢行系统的构建，以连贯、安全、舒适、丰富功能为目标。设计对场地内选为绿道的几条街道，提出了具体的改造策略：对于连接元大都与转河的西直门北大街，利用高架桥桥下空间，改造交叉口；设计自行车道及路侧停车；人群聚集处减少停车位，改为自行车停车。对于学院南路，设置自行车道，与停车位之间设置1m缓冲带；有底商处，可取消2~4个停车位，设置街边"微公园"，提升场地活力。对于文慧园北路，将原来的双侧停车改造为单侧双向自行车道；有底商处设置雨水花园，完善生态功能。对于余地较小的道路，采用垂直绿化方式，地面设计不同颜色的铺装线，以提示附近公园及绿地。

　　对于街旁绿地与小区游园的改造，结合场地现状，适当拆改质量较差的建筑、平房等，增加休息座椅及活动空间，丰富社区功能。最终形成完善的社区绿色微循环体系。

　　重点更新地段位于学院南路32号院内。该居住区被一条城市支路分成两半，支路两侧多棚户建筑，主要为餐饮、KTV等商业内容。设计以低成本、丰富功能、方便居民为更新策略，增加老年人活动空间，引入城市农业；引入成本低、组装灵活和形式丰富的集装箱建筑；迁入附近原有菜市场，方便周边居民日常生活。设计遵循可持续原则，利用回收的海运集装箱，注重自然采光及通风，设计保温屋顶，局部设置太阳能板、绿墙及屋顶绿化，在有限的空间增加绿化率，以创造最大的生态效益。注重居民参与，征集居民意见。最终形成集商业、居住、休闲于一体的社区公园。

规划设计 PLANNING & DESIGN

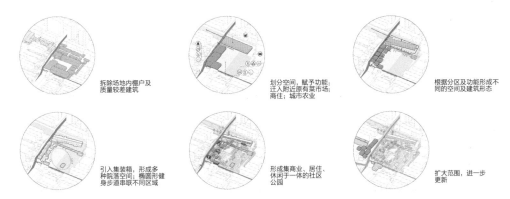

更新策略
Renewal Strategy

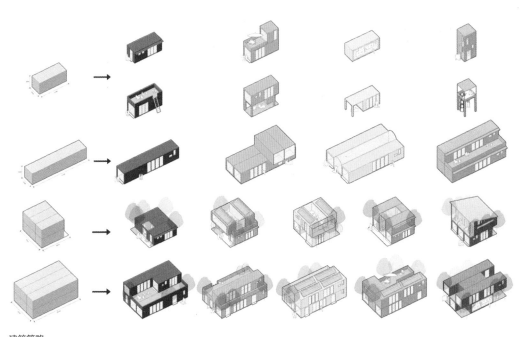

建筑策略
Architecture Strategy

街区效果
Perspective of Neighborhood

喷泉广场效果
Perspective of Fountain Square

6. 人定湖公园周边改造提升

场地位于北京市西城区、海淀区交汇处，在上位规划中功能定位属于邻里休闲区。人定湖公园是一个以游憩、穿行为主要功能的社区公园。其活力与周边业态、居住区息息相关，在设计时应尽可能多地保留现状的居民区和业态，不打破原有的生活平衡，在此基础上进行绿网的构建。

根据区域策略，人定湖公园是区域内步行核心，规划打通断行路网，规划绿网骨架，使慢行体系通畅。同时开辟公共绿地，现状公园作为区域绿心，街旁绿地作为街区绿核。最终形成南北贯通的绿网体系。

分析场地内部现状问题。对占道停车、街区立面不美观、街道绿化风貌单一这三个主要问题，提出不同的解决策略。

为增加绿网慢行舒适度，改善街道不美观、脏乱等问题，在道路界面改造方面，针对墙体、开放居民楼和商铺三种类型界面提出更新策略。墙体界面，在有余地处种植乔灌木，余地小处增加垂直绿化；开放居民楼界面，将灰空间整合为自行车停放、垃圾收集等功能条带，使街道功能集约畅通；商铺界面，拆除临时棚户、重建立面，让出新的绿地空间，在部分商铺前可采用盆栽等可移动绿化增加绿色界面。

为使道路绿化更加丰富，改善绿化风貌单一的问题，对现有绿地进行局部调整：绿化功能带调整，将公车站和绿带结合，使人行道更畅通；绿化现状改造；重点绿化节点强化，在商业、历史文化等界面种植符合该场地氛围的植物种类。

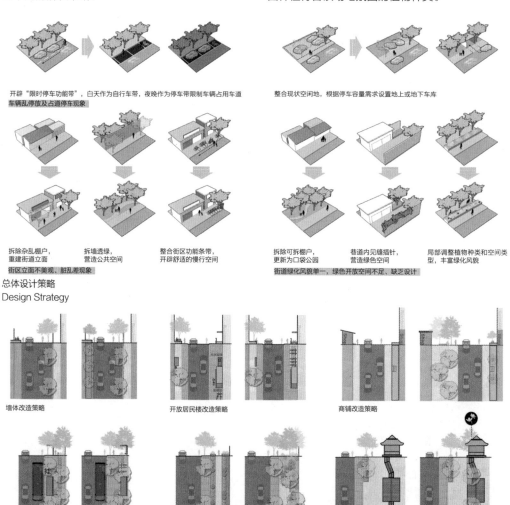

总体设计策略
Design Strategy

道路界面改造策略
Strategies of Road Border Transform

规划设计 PLANNING & DESIGN

在重点地块设计方面，主要对人定湖北巷及周边区域进行了重新规划。交通承接上位规划，将原有道路进行南北向打通。场地北部保留原有居民业态，将现有的杂乱商铺整合到一栋集市楼内，利用空余用地创造更多绿色空间；中部雨水花园处理市场废水，同时起到绿化过渡作用；南部开放原有的封闭运动场，并将运动场南侧打造成运动公园，形成一个具有运动功能的公共绿地，增加周边居民的活动场地，一定程度缓解人定湖高峰时段的人流压力。

位于五通路的街旁花园，结合周边的小学和社区，利用现状空闲地设计兼具地库功能的口袋公园，缓解停车问题的同时提供了新的开放空间。

位于安德里北路街口处的空闲地，将原有卫生站转移，结合地库及穿行需求设计为街旁绿地，打开了路口空间。

通过对人定湖周边区域的道路及重点地块的改造设计，最终形成该区的主要绿网与绿色节点，满足周边居民的日常休闲生活，完善区域内慢行体系的构建。

公园入口效果
Perspective of Sport Park Entrance

运动公园效果
Perspective of Sport Park

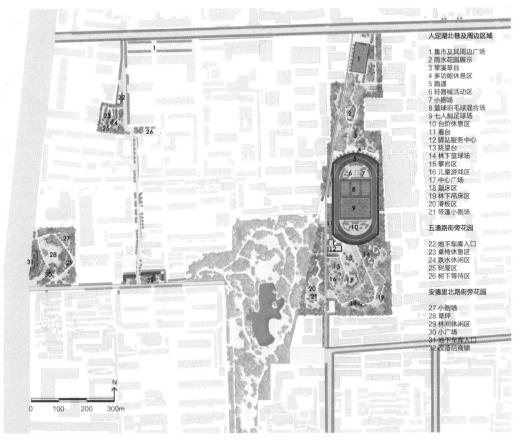

人定湖周边地块设计平面
Design Plan of Rendinghu Area

The site is located on the north of Beijing Second Ring Road and the west of the central axis of the city, which is the junction area of Xicheng, Haidian, and Chaoyang District. This site is of great significance for the continuation and connection of urban green space.

The site is mainly for residential use, and the internal facilities are perfect. The function of residence and office can be met, but it is too closed. The public service facilities are hard to serve the surrounding residents; the commercial land is scattered and there is no regional commercial center; the green space and public space are inadequate , and there is a lack of connection between them, therefore the overall benefits cannot be achieved. The residential land is mostly isolated by walls, and although it has attached to the green areas, it lacks connection with the surrounding areas and the utilization rate is low. As to the traffic land. The road network in this area basically met the needs of the crowd, but there are problems in the optimization of the route. The network of road is not perfect, and the urban road network is incomplete and there are several broken roads. Due to the existence of broken roads and the walls of courtyards, the accessibility of slow traffic systems is extremely poor. The traffic of Deshengmen roundabout is more chaotic, Which is a mix of pedestrian and motor traffic. The parking lot is fragmented. Although there are a large number of underground parking lots, it still cannot meet the parking requirements in the whole area. Many streets are full of vehicles on both sides, occupying public space and causing road congestion.

The goal of this plan is to restore the original natural base and build the urban green ecological network. In details, it aims to stimulate the vitality point, to activate the commercial cultural facilities, to improve the vitality of the lot, to make the urban green space give full play to its role of the ecological efficiency , and to ensure the sustainable development of the city. A methodological system for ecological network planning of urban green areas based on suitability analysis will lead to distinct planning approaches, objectives and methods, and better organized planning process as well. The planning results will be more scientific.

The planning strategy selects the factors that best reflect the suitability of the urban ecological network. Decision analysis, subjective judgement will be based on the expertise of many subjects such as urban planning, ecology, hydrology, geology. After several rounds of factor selection-evaluation-determining weights-superimposing calculations, the most representative and most reflective ground-basis factor is identified. As a representative of the suitability of greenbelt ecological networks, a single factor suitability map is created. The larger the assignment, the higher its ecological sensitivity. The network is based on the two basic units of plaque and corridors, which are organized into a whole by structure. The "stream" movements of materials, energy, and information are the basic functions of the network. The urban green area ecological network also has the general characteristics of the network, which is reflecting a spatial connection model of the surface landscape. The ecological network consists of patches and corridors in its spatial structure. Gravity through the ecological network planning - the method of resistance analysis can help to determine the construction of the network. The resistance value of the calculation model is mainly used to determine the direction of the corridor, and to ensure that the corridor receives the least resistance on the landscape. On this basis, nodes and corridors are identified and superimposed to form the ecological network structure of the land.

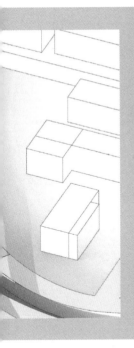

蓝绿交织
基于适宜性分析的城市生态网络构建

Intertwining Blue and Green
Construction of Urban Ecological Network Based on Suitability Analysis

刘涵、高敏、阎姝伊、刘喆、胡凯富、刘峥
Liu Han / Gao Min / Yan Shuyi / Liu Zhe / Hu Kaifu / Liu Zheng

场地位于北京市二环路以北、中轴线以西、三个城区的交界地带，地势平坦。西至西直门大街，南至德胜门西大街、北二环，东至鼓楼外大街，北沿新康路、黄寺大街，总面积约为46hm^2。场地内开放活动空间丰富，具有人定湖公园、北滨河公园以及沿河绿带等多处绿地，同时具有德胜门、西黄寺等历史文化节点。但场地内同时存在诸如绿地分布分散、交通不连贯、景观结构不合理等一系列问题，需要通过合理的场地规划实现居民对绿色开放空间的最大化利用。

此次规划以城市双修政策为大背景，力求建立基于适宜性分析的城市绿地生态网络规划的方法体系，使规划的途径、目标、方法更加清晰，规划过程更加具有条理，规划成果更具科学性，打造真正的城市生态网络。

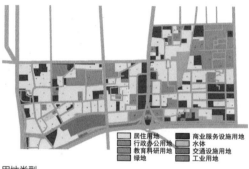

用地类型
Land Type

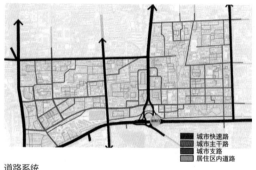

道路系统
Road System

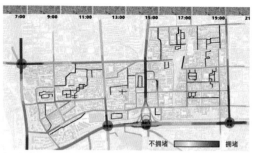

交通拥堵程度
Traffic Congestion

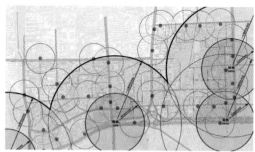

公共交通服务半径
Radius of Public Transport Services

道路可达性
Road Accessibility

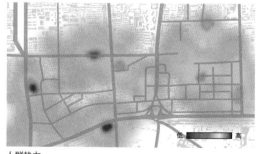

人群热力
Crowd Heat

建筑年代
Construction Age

建筑售价
Construction Price

居住用地是该地块主要用地类型；行政办公用地，内部设施完善，居住、办公等功能均可满足，但却过于封闭，公共服务设施难以服务周边居民；商业用地零散、无区域性商业中心。整体地块中缺乏绿地和公共空间，且地块之间缺乏联系，无法发挥整体效益，由于居住区多、多为小区级绿地和公共服务设施，与城市整体协调性不够，与周边地区发展不协调；交通用地，车辆段功能外迁，可考虑拆除。

该区域中路网已基本满足了人群使用的需求，但是在路径的优化上存在问题。路网体系不够完善，城市道路网络残缺、存在断头路，加上大院围墙，慢行系统可达性极差。德胜门环岛交通混乱，缺少人车分流。停车场分布分散，虽然有较多数量的地下停车场，但依旧不能满足区域内的停车要求，不少街道两侧停满车辆，占用公共空间，造成路段局部拥堵。城市快速路在工作日早晚高峰会出现拥挤现象，局部会严重拥挤。局部道路也存在由于道路等级不能满足车流量需求而出现拥堵的情况。区块内大多建筑的建设年代集中在1996年以前，建设时期还流行大院单位制，导致产生了很多大院。

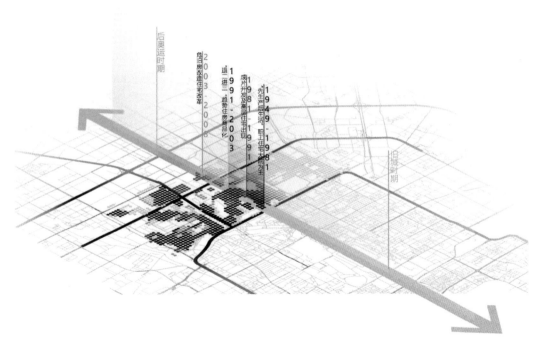

住区发展时序
Time Series of Residential Area Development

绿色空间统计
Green Space Statistics

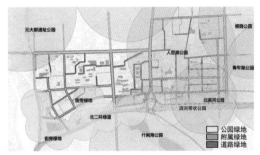

绿地类型和服务半径
Green Type and Green Service Radius

规划目标：恢复自然基底，并在此基础上构建城市绿色生态网络；激发活力点，激活商业文化设施，为地区带来生产生活动力；带动城市双修，恢复地区活力，利于城市绿地发挥的生态效能，保障城市可持续发展；建立基于适宜性分析的城市绿地生态网络规划的方法体系，使规划的途径、目标、方法更加清晰，规划过程更加条理，规划成果更具科学性。

规划策略：选取体现地块城市生态网络适宜性特点的水因子等，具有翔实资料且可利用程度高的建筑、绿地等因子。基于多种学科的专业经验，根据专家决策法进行因子的筛选及评价，综合多种学科背景的专家，如城市规划、生态学、水文、地质等进行决策分析。经过多轮的因子选择—评价—确定权重—叠加计算等过程，确定最具代表性的基础因子，作为绿地生态网络适宜性的代表，其赋值越大代表其生境越差，以此来形成单因子适宜性图。网络基于斑块和廊道这两个基本单元组成，两者通过结构组织成为一个整体。物质、能量、信息等"流"的运动是网络的基本功能。城市绿地生态网络同样具有网络的一般性特征，反映了地表景观的一种空间联系模式。生态网络在空间结构上由斑块和廊道构成。通过生态网络规划引力－阻力分析的方法构建网络。阻力值计算模型主要用来确定廊道的走向和方向，确保廊道在景观面上所受的阻力最小。在最终的生态网络基础上，对节点和廊道进行识别和叠加，形成规划平面图，通过对场地的详细设计和分析、反馈和修改，形成最终的规划设计平面图。

绿地利用潜在地块
Potential Plots

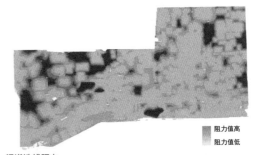

绿道选线阻力
Resistance of Greenway Line Selection

生活绿道
Life Greenway

通勤绿道
Commute Greenway

健身绿道
Fitness Greenway

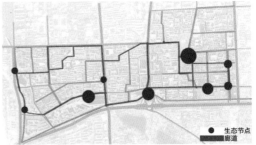

生态节点与绿道的识别
Ecological Node and Greenway

规划设计 PLANNING & DESIGN   37

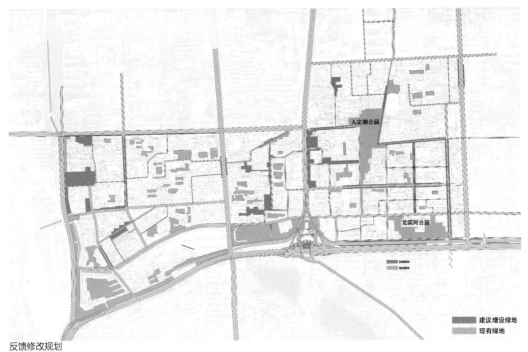

反馈修改规划
Feedback Revision Plan

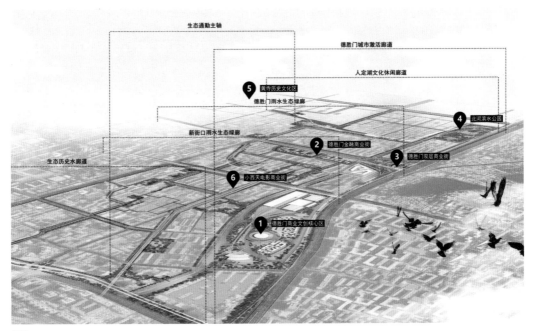

鸟瞰
Aerial View

规划设计 PLANNING & DESIGN

规划总平面
Master Plan

规划设计 PLANNING & DESIGN

1. 社区开放空间改造

场地位于北二环边，北至学院南路，南至北护城河边，西至西直门北大街，东临文慧园北路。西直门北大街是连接南北向的城市主干道；学院南路是场地内东西向的主干道，次干路为文慧园北路、联慧路和文慧园路。场地人群主要分布于商业、十字路口、绿地周边。场地周边公共交通站点较多，分布均匀，能够满足场地内部人群的出行需求。但由于换乘，导致外来人群在通勤期间穿越场地频率较多，对场地交通造成影响。场地内现状主要是以居住和学校用地为主。大院与居住区的围墙将场地与外界隔离，仅通过几条道路联通其他区域。居住区大院现状以封闭围墙围合，居住区中场地的使用人群主要以老人、儿童为主；大院内建筑包含居住和办公功能，虽满足自身生活需求，但影响周边用地的通行和使用。

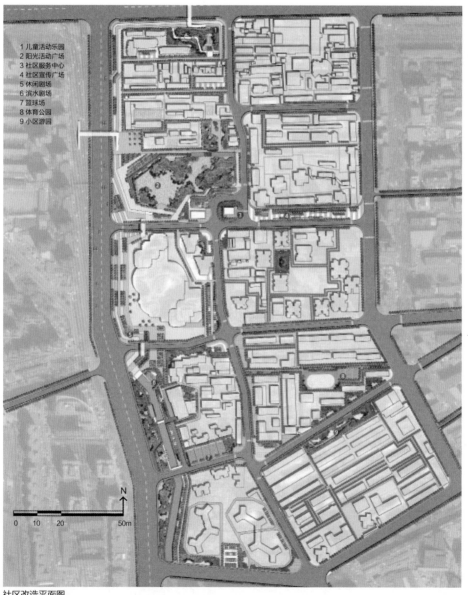

1 儿童活动乐园
2 阳光活动广场
3 社区服务中心
4 社区宣传广场
5 休闲剧场
6 滨水剧场
7 篮球场
8 体育公园
9 小区游园

社区改造平面图
Design Plan of Neighborhood Renewal

规划设计 PLANNING & DESIGN

潜在绿地分布
Potential Green Distribution

道路系统
Road system

人群热力分布
Crowd Heat Distribution

用地类型
Land Type

学院南路段现状
Existing Condition of College South

学院南路段改造效果
Perspective of College South

枫蓝国际段现状
Existing Condition of Maple Blue International

枫蓝国际段改造效果
Perspective of Maple Blue International

转河段现状
Existing Condition of Zhuanhe

转河段改造效果
Perspective of Zhuanhe

2. 建筑综合体设计

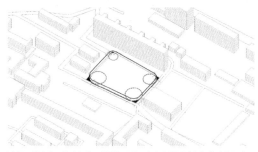

选择区域内连通性良好的区域进行设计。考虑人流活动行为等诸多因素，对建筑功能进行定位

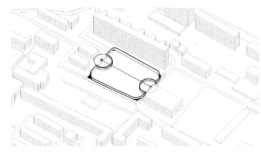

庞大的建筑体量会对市民日常生活造成诸多影响，利用两个大型的圆弧广场来削弱建筑体量感，带入人流

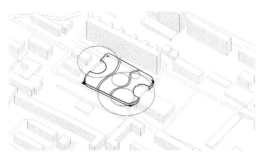

考虑到一层部分的基本光照需求，在体量中减出弧形的天窗采光

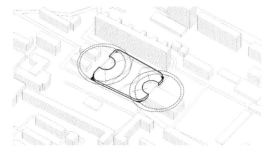

大体量的屋面公园需要制高点来构筑建筑的基本形式，由弧形广场向建筑体量内化的延伸，产生弧形天桥元素

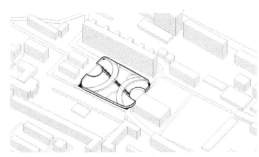

建筑自身功能层次的分割促使各层间需要特别的交互，在整个体量中心进行斜向切割，构筑下沉的楼梯空间

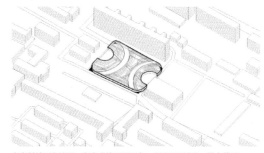

参考基地原有地形，考虑建筑屋面汇水，设计屋面地形，覆土草顶

地形低洼处形成汇水井，井中设立雨水管道，在井中种植树木植被

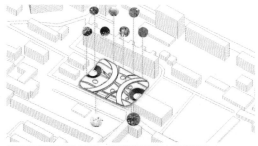

增加屋面公园植被的多样性，满足整体美观和人们日常休憩

建筑综合体生成过程
The Formation of the Building Complex

规划设计 PLANNING & DESIGN   43

屋面成为市民日常休憩、活动的公园，一层为商业、体育活动的场所，底层作为地下车库使用。

地下车库一层将入口与城市干道相接，在建筑前方下沉绿地形成回车场地，并增加绿化、健身场所等诸多功能，出口面向机动车通行较流畅的文慧园路，一定程度疏散了机动车拥堵情况。建筑首层作为商业使用，在前期的调研中发现，周边场地的商业分散且混乱，集中的商业能很好地促进居住区的商业气氛，改善市民生活。

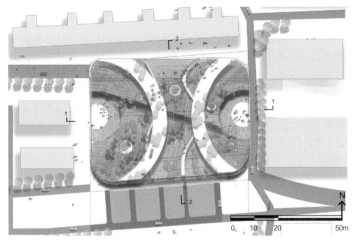

建筑综合体设计平面
Design Plan of Building Complex

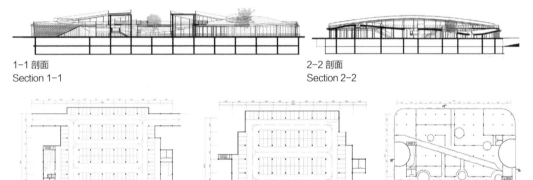

1-1 剖面
Section 1-1

2-2 剖面
Section 2-2

分层建筑平面
Building Layout

鸟瞰
Aerial View

3. 德胜门立交桥下沉公园

德胜门立交桥复杂的交通状况和交错的现状条件是改造的重点所在，现状中车行交通、公共交通和人行交通穿插，水系与德胜门互相交错，改造的重点在于整理场地内部的交通流线，利用现在场地内部的水系和古建来营造适宜的公共空间。

改造后将德胜门前后的公交站点移走，并结合西北护城河的水系打造德胜门下穿公园，将人行和自行车行引导至德胜门下，与机动车分隔。

德胜门立交桥下沉公园由德胜门北侧的跌水下沉公园和德胜门南侧的回望平台构成，北侧着重解决交通问题，设置多个出口分别连接了德胜门、公交站点、公交站和人行天桥。利用高差塑造多种类型的景观。

南侧坡度较缓，利用原来二环路的天窗上盖，塑造平远的草坡地形，将德胜门改造成区域性的标志景观，将大片的城市空闲地塑造成可供行人游客停留、通行的绿地，连接上盖所覆盖的街区和公交站点。

机动车行流线分析
Analysis of Vehicle

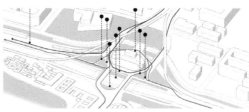

公共交通及站点分析
Analysis of Public Transportation and Bus Stop

人行交通流线分析
Analysis of Pedestrian

水系分析
Analysis of Water System

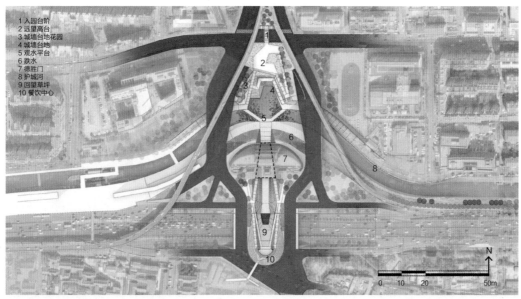

德胜门改造设计平面
Design Plan of Deshengmen

规划设计 PLANNING & DESIGN

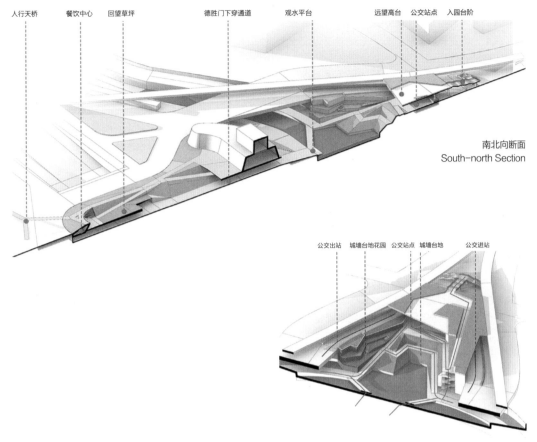

南北向断面
South-north Section

东西向断面
East-west Section

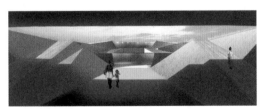

下穿隧道效果
Perspective of Tunnel

城墙台地效果
Perspective of Wall Werrace

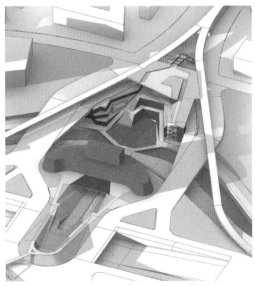

鸟瞰
Aerial View

4. 旧城区邻里空间更新

邻里社区微更新段的规划目标是通过建设一个公园，焕活社区居民日常休闲娱乐的生活方式；一条绿道，开辟居民社区健康出行的绿色通道；几处绿地，提供社区居民"所想即所得"交流场所。老旧小区的空间施行"微更新"策略，通过拆改棚户区、老旧危房，用细小的"空间手术刀"治理社区问题，力图为社区开辟新的生活篇章，打通微循环，焕活社区动力，实现社区更新。设计区域拥挤、老化、缺少活力、缺乏公共空间。因此，对关键空间实行"微提升"，是解决老城区社区问题的一种途径。

首先对社区交通进行考察和新的微规划，用最小的干扰方式，打通社区多条断头路，连接邻里空间，串联家家户户，为居民到达规划绿地提供出行条件。因此，打通微循环是规划的第一步。

其次，对社区内的绿地进行梳理，对小片的不成体系的绿色空间进行整合，将功能赋予场所，以便居民使用。通过社区沟通，对居民的愿望进行梳理统计，作为设计的重要参考，用不同的空间承担不同的使用功能，自然形成微干扰的公共空间体系。

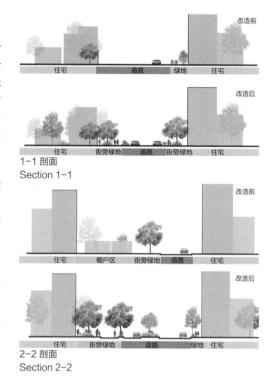

1-1 剖面
Section 1-1

2-2 剖面
Section 2-2

旧城区设计平面
Design Plan of Old Areas

规划设计 PLANNING & DESIGN

喷泉广场效果
Perspective of Fountain Square

社区游园入口效果
Perspective of Garden Entrance

雕塑广场效果
Perspective of Sculpture Square

街心公园效果
Perspective of Street Garden

绿道效果
Perspective of Greenway

5. 人定湖西侧周边环境提升

区域以人定湖公园为绿心，此区域周边交通发达，周围用地主要以老式居住区为主。设计目标是将周围绿地串联成绿廊，打造休闲舒适的大型生活居住环境。方案由德胜门外大街附近的开放空间、居住区游园、社区生活步行街及地下停车位、绿道建设四方面组成。

德胜门附近，在街旁公共开放空间增设周末集市与小型草坪剧场以满足人们游览需求，在医院旁设计冥想花园、康复花园、芳香园帮助病人恢复健康，为不同人群提供多种服务。

在居住区，拆除原有场地的棚户区，整合形成较为完整的空地，结合居民使用状况，增设多样的运动休闲活动设施。

在社区内整合部分餐饮等一系列基础设施，打造社区生活步行街，满足人们买菜、吃饭、休闲娱乐一系列活动，为解决步行街两侧居民的日常出行，增设地下交通保证地面、地下功能互不干扰。

由于人定湖公园内仅允许步行通过，南北向交通受到阻隔，新增道路保证南北向连通性，将原有被阻隔的道路一并打开，保证路网的连通性和完整性。

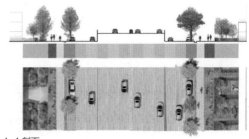

1-1 剖面
Section 1-1

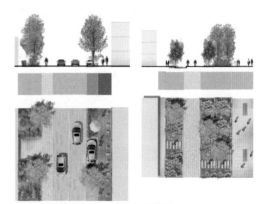

2-2 剖面
Section 2-2

3-3 剖面
Section 3-3

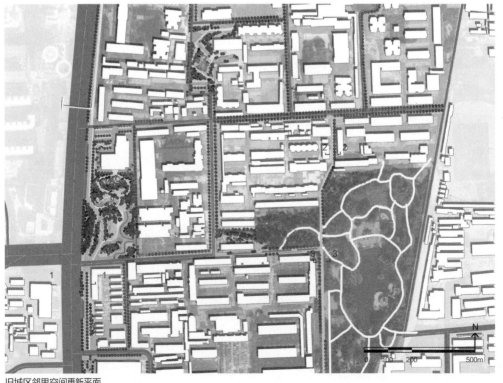

旧城区邻里空间更新平面
Design Plan of Historical Area Renewal

规划设计 PLANNING & DESIGN

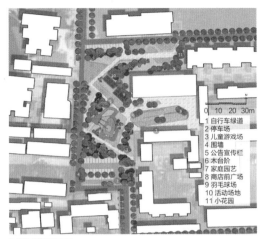

居住区游园平面
Design Plan of Neighborhood Garden

1 自行车绿道
2 停车场
3 儿童游戏场
4 围墙
5 公告宣传栏
6 木台阶
7 家庭园艺
8 商店前广场
9 羽毛球场
10 活动场地
11 小花园

居住区游园效果
Perspective of Neighborhood Garden

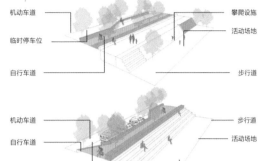

德胜门外大街旁开放空间平面
Design Plan of Deshengmen Open Space

1 自行车绿道
2 临时停车位
3 港湾式公交站
4 树冠桥
5 草坪剧场
6 活动场地
7 冥想花园
8 体验花园
9 芳香道
10 停车场
11 周末集市
12 舞动广场

德胜门外大街旁边界空间分析
Analysis of Deshengmen Border Space

德胜门外大街旁开放空间效果
Perspective of Deshengmen Open Space

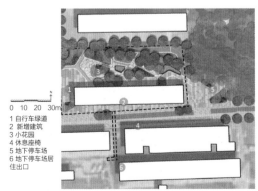

社区生活步行街及地下停车场平面
Design Plan of Pedestrian Street

1 自行车绿道
2 新增建筑
3 小花园
4 休息座椅
5 地下停车场
6 地下停车场居住出口

社区生活步行街效果
Perspective of Pedestrian Street

6. 人定湖东侧设计

地块位于西城区安德里北街附近，场地周围主要是居民区以及沿街商业区。现状绿地分散且有多处闲置空地，临近人定湖有一片废旧建筑，整体效果不佳，且不具有保护意义，建议拆除。由于该地块社区集中且封闭性强，造成环境和开放空间的割裂。因此，设计核心主要考虑如何打开及连接各个居住区来创建开放性绿色社区。

场地内主要以居住用地及行政办公用地为主，由于行政办公用地内部主要为军队大院，保密性强，因此设计上考虑保留原状。局部分布有小学及医院，场地西侧紧邻人定湖公园。

设计分为三个地块，主要从公共空间、经济效益、格局改造、安全问题和生态网络五个方面为切入点开展设计。

德胜门外大街邻里空间保留了原操场，拆除周边质量较差的平房建筑，设计一块以体育锻炼为主题的街旁绿地，与人定湖公园绿廊相连，功能互补，服务于周边人群，形成楔形绿廊。德胜门广场位于北京中轴线上，该街区人流量较大，但缺乏一块行人停留空间，因此将该地块沿街打开，打造服务于周边人群的城市开放空间。德胜门建筑附属空间位于西黄寺对面，设计拆除部分沿街餐饮建筑，打造一块具有展览体验功能的街头广场，运用地雕、展廊等元素体验佛学文化。

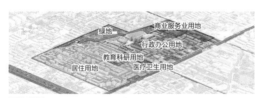

用地分析 Land use

道路分析 Road Analysis

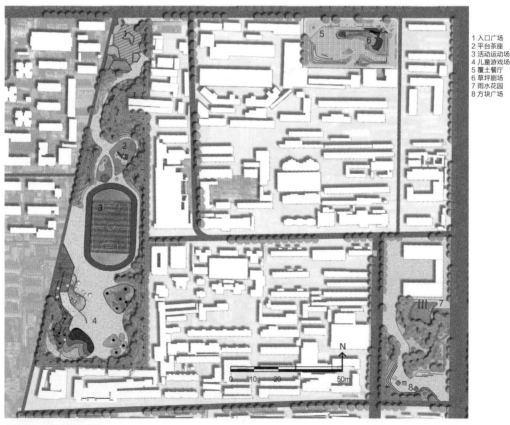

1 入口广场
2 平台茶座
3 活动运动场
4 儿童游戏场
5 覆土餐厅
6 草坪剧场
7 雨水花园
8 方块广场

人定湖东侧提升改造平面
Design Plan of Rendinghu East Area

规划设计 PLANNING & DESIGN

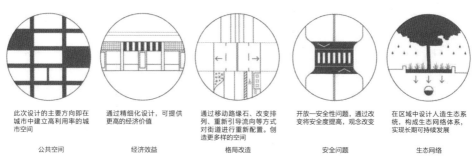

| 公共空间 | 经济效益 | 格局改造 | 安全问题 | 生态网络 |

此次设计的主要方向即在城市中建立高利用率的城市空间

通过精细化设计，可提供更高的经济价值

通过移动路缘石、改变排列、重新引导流向等方式对街道进行重新配置。创造更多样的空间

开放—安全性问题，通过改变将安全度提高，观念改变

在区域中设计人造生态系统，构成生态网络体系，实现长期可持续发展

设计概念
Design Concept

德胜门广场现状问题
Current Situation

德胜门外大街邻里空间现状问题
Current Situation

德胜门建筑附属空间现状问题
Current Situation

德胜门外大街邻里空间效果
Perspective of Deshengmen Neighborhood Space

德胜门广场效果
Perspective of Deshengmen Square

德胜门建筑附属空间效果
Perspective of Deshengmen Architecture Space

德胜门建筑附属空间效果
Perspective of Deshengmen Architecture Space

The site is located on the east side of the north central axis of Beijing. Beside the site, Xiba River Green is on its east and the Moat Green of the Second Belt is on its south. Ditan park, Youth Lake Park and Willow Park are the most remarkable parks in this area. Meanwhile, the Moat Green of the Second Belt and Xiba River Green are two of the green ways in city level which play a vital part in constructing the green way network of Beijing City. In addition, the historical and cultural tour of the ancient city and education also play important roles in this area as it closes to the north central axis of Beijing.

In the early days of the founding of the People's Republic of China, this area was basically a field of wild land scattered with mud ponds and weeds outside the Ditan Park. During the socialist construction period in the 1960s, Youth Lake Park and Willow Park were constructed as examples. Besides, many factories and governmental units were built in this area, which was the basic building form ,and then were turned into residential districts years later. There is plenty of space for improvement in the quality of living environment as it was constructed long time ago and the facilities were rusted out now. The structure of Three Cores with an Axis, which means that three main parks are recognized as the cores and connect the two green ways and the north central axis of Beijing, has been formed as the structure of green space in this area were planned well. However, green space in this residential district was separated and deserted. Although the transportation system is relatively complete and convenient, the slow traffic is inadequate because of the low-quality landscape along street, old and rusty facilities. Also, it is significant to connect the historical and cultural relics inside and outside this area in the design plan.

Generally, on the one hand, this site is rich in resources and potential of development; on the other hand, the whole area needs to be activated and renewed immediately since the quality and animation of the green space is inadequate and insufficient.

The main purpose of planning and design is to superimpose green space and other urban functions through the concept of "green space+", using complex green spaces and greenways to renew old city distric, to strengthen historical and cultural landscapes, and to enrich modern urban life. Finally, the construction of urban green network will activate the old urban area.

Thus, the following planning steps are proposed: Firstly, the functions of the site are divided into three aspects: ecology, history and living. Secondly, potential land plot selection and greenway selection are conducted based on site analysis. Finally, three functional areas including park recreation areas, cultural experience areas, living and leisure areas are formed, which are connected by three types of greenways.

Specifically, Liuyin Park and Youth Lake Park are the cores of the park recreation area. The Ditan Park and its surrounding historical sites form the cultural experience area. The remaining courtyards and residential areas form the living and leisure area. According to the functional characteristics of these areas, various sites have been specifically designed and modified. The transformation strategies of park recreation areas are mainly to open park boundaries, add sports parks, and increase community parks. The strategies of the cultural experience area are to inject the cultural and creative industries, add creative markets, and hold outdoor exhibitions. The strategies of living and leisure areas are mainly to create ecological roof gardens, demolish community walls, and integrate broken public spaces. Based on the above strategies, commercial transformation strategies will be conducted to activate the potential land and meet the daily activities of the residents. In addition, links will be provided by designated greenways to improve public service facilities, renovate the streetscape of the greenways, and provide high-quality walking and cycling experiences.

In the end, the renovation of the green network based on the conception of "green space+"will promote the renovation of the city and the restoration of the ecological environment, and integrate urban functions, complete a slow-traffic system, and improve the ecological environment.

"绿地+"
基于复合型绿地整合的北中轴绿网更新

Green Space +
The Renewal of Beijing Central Green Network Based on Green Space Integration

王思杰、蒋鑫、宋怡、邢露露、姜雪琳、韩炜杰、张希
Wang Sijie / Jiang Xin / Song Yi / Xing Lulu / Jiang Xuelin / Han Weijie / Zhang Xi

地坛始建于明代嘉靖九年（公元1530年），是明清两朝帝王祭祀"皇地祇神"的场所，距今已有488年的历史。其所在片区的发展是北京由历史古城向现代都市转变的缩影，具有独特的研究价值。

场地位于北中轴的东侧，二环护城河的北侧，既是重要的历史文化节点，也是亟待整治更新的老旧生活片区。场地内部绿地量大但连接度低，区位定位与其现有功能不匹配，现有绿地不能满足丰富的城市现代生活。

此次规划希望在更新旧城绿网，改善绿地的同时，赋予绿地更丰富的内涵；依托现有绿地空间，串通绿色廊道，活化社区微循环；以空间开发为导向，以绿色空间为介质，将绿地与文创、商业、生态功能相融合，激发场地潜力，打造功能复合的活力绿色网络。

规划区域位于北京北二环与北四环之间的北京市北中轴两侧，是北京市的中心区域，南侧连接北京老城区，北侧连接奥林匹克公园，东西两侧邻接文创产业区和海淀教育与高科技园区。

规划区域在北京绿地系统中的位置十分重要，处于北边和西北边五条绿楔与北京老城区护城河绿带的连接处，并且与五条规划绿道相接，绿地的连接、整合潜力非常大。该地区在新中国成立初期基本为泥塘与野草的城外野地，20世纪60年代社会主义建设时期开挖了青年湖和柳荫湖，并在此处设置了多处厂房和单位大院，形成了此区域的基底。

改造区域的绿地率为45%，人均绿地面积为37.22m²/人，远高于东城区12.07m²/人的平均水平。整体来看，西侧绿地结构虽然较为完善，但连接性差，东侧居住区绿地较为分散，绿量不足且不成体系。

规划区域各类问题存在的根本原因是该区域在新中国成立初期时只是作为城市边缘的大院来规划的，而现在该区域已成为平均房价极高的城市中心区域。随着时代的变化，人们的需求也发生了较大改变，该区域应当具有与现状相配套的居住环境。

规划愿景是借助功能复合型绿地带动整片区域的绿网更新，同时也提高了该区域在商业服务、历史文创、体育健身等多方面的水准，让这片区域成为既有良好的绿色生活环境、又有丰富的城市现代生活的充满活力的中心城区。

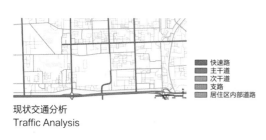

现状交通分析
Traffic Analysis

现状绿地分析
Green Space Analysis

更新旧城绿色网络
Update the Green Network of Old City

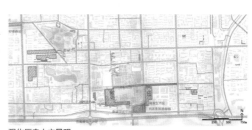

强化历史人文景观
Strengthen the Historical and Cultural Landscape

丰富城市现代生活
Improve Urban Modern Life

最终潜力地块选择
Select the Final Potential Sites

潜力地块选择
The Selection of Potential Sites

规划设计 PLANNING & DESIGN 55

生态休闲类绿道选线
The Ecological and Recreational Greenway

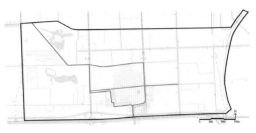

历史文化类绿道选线
The Historical and Cultural Greenway

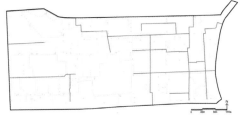

邻里生活类绿道选线
The Neighborhood Living Greenway

最终绿道选线
The Final Greenway

绿道选线
The Selection of Greenway

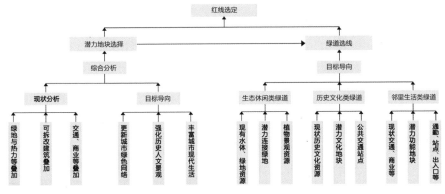

红线确定策略
Strategy

　　红线确定的基本策略是：先由对现状的分析结合规划目标综合分析，选出该区域的潜力地块；再依据现状绿地、历史文化、交通等资源，结合三个目标导向，用三类绿道把属于各自类型的潜力地块连接起来，构建出符合规划目标的多功能绿网。

　　潜力地块的选择基于目标导向性，通过叠加人口热力图、现状绿地及基础设施、经评估后的可拆除建筑等因素得到最终结果，以满足更新城市绿色网络、强化历史人文景观和丰富城市现代生活的功能需求。

　　绿道选线也是基于目标导向型，分别得到生态休闲类绿道、历史文化类绿道、邻里生活类绿道。

　　生态休闲类绿道通过叠加现状水体、现状绿地和潜在绿地因素得到；历史文化类绿道通过叠加历史遗迹、相关文化景点、潜力地块和公共交通因素得到；邻里生活类绿道通过叠加居住片区、社区围墙及其出入口、公共交通、学校分布、商业分布、现状绿地分布及潜力地块因素得到。

　　最后，现状绿地、潜力地块以及绿道选线相叠加得到最终的红线范围。

规划设计 PLANNING & DESIGN

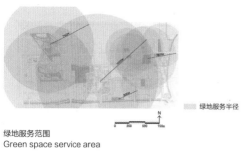

绿地服务范围
Green space service area

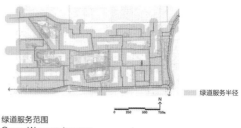

绿道服务范围
Green Way service area

规划分区
Planning Zoning

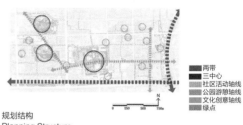

规划结构
Planning Structure

规划分析
Planning Analysis

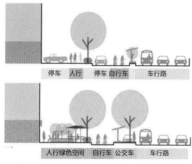

道路类型 1
Road Type One

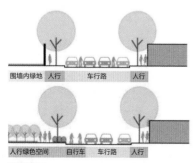

道路类型 2
Road Type Two

路径处理模式
Path Design Mode

所选绿地服务范围基本能够完全覆盖整个片区，绿道的服务范围基本能覆盖本片区 94% 的社区、公园、学校等公共服务设施的出入口。最终将整个规划的绿网体系分为公共游憩区、文化体验区和生活文化区。

将所选绿道按等级可以分为三类：城市主干道、次干道、支路。然后针对不同的绿道类型提出了相应的改造策略，基本目标是提高安全性，创建更加开放的步行空间，形成舒适的城市慢行体系。

城市主干道现状问题是大量路侧停车位使得自行车道变窄，危险性增加，并且路侧建筑附属空间未能充分利用。改造策略是调整道路断面的划分，增设道路停车绿岛，并将自行车道与机动车道隔开，提高安全性。路侧建筑附属空间打开，形成开放的人行绿色空间。

城市次干道现状问题是存在人车混行情况，严重影响行车安全；社区围墙封闭性高，缺乏内外联系。改造策略是调整车道划分，确保人车分流；将社区围墙拆除，释放部分空间作为城市慢行体系空间，增加社区与城市的联系性，改善步行体验。

城市支路现状问题与次干道类似，由于道路狭窄，空间不足，人车混行状况更为严重。改造策略是打开围墙，释放空间，同时增设一定的分车带，在保证机动车行驶的基础上，为行人提供舒适宜人的绿色步行空间。

该片区大面积的公园和绿地在中心城区的存在是十分珍贵的，但绿地并没有得到充分的开发和利用，绿地没有产生与它所在土地匹配的价值和效益。绿地是人们愿意停留的场所，植入消费、健身等其他功能，绿地的利用率和效益就提高了。规划中赋予绿地更多的功能，"绿地＋商业"产生经济效益；"绿地＋历史文创"产生文化效益；"绿地＋体育健身"产生社会效益。鼓励商业、文创、体育等功能的入驻，提高绿地人气，有人气之后，绿网的维护、交通等问题便迎刃而解，借助市场和社会的力量完成整个绿网更新，实现规划目标。同时也提高了该区域在商业服务、历史文创、体育健身等多方面的水准，让这片区域成为既有良好的绿色生活环境、又有丰富的城市现代生活的充满活力的中心城区。

首先，该区域三个最大的公园都集中在此，连接潜力巨大。为了提高三个公园各自间的连接性，着重设计了两个连接节点。同时，根据改造后的公园出入口和边界，梳理出一条串联三个公园绿地的环线，使三个公园连接成一个体系。环线道路通过铺装和标识系统的设计，引导居民和游客进入公园活动。

其次为各个社区部分。首先是街道绿色空间体系的整改。由西侧三大公园向东侧延伸至各个社区组团，进一步串联各个社区组团，形成整改后的街道绿色空间体系。城市级的街道绿色空间整改分为两种，一种是围墙阻隔造成建筑附属空间的浪费，策略是拆除围墙，构建绿色步行空间；另一种是路侧停车占道及人车混行问题，策略是调整车道并增设路侧停车位。社区级的街道绿色空间整改也分两种，一种是改造或打开围墙增加开发空间；另一种是增设生态停车位以解决停车占道问题。

最后进行廊道的串联，场地现有的廊道主要有二环滨河绿带和东侧的西坝河绿带，在此基础上，增加了城市东西向与南北向的线性空间，完成整体绿网更新。

最终实现规划愿景，基于"绿地＋"的理念，以绿色网络的更新带动城市功能修补和生态环境修复，构建融合城市功能、完善慢行体系、注重生态环境的具有综合效益的绿色网络体系。

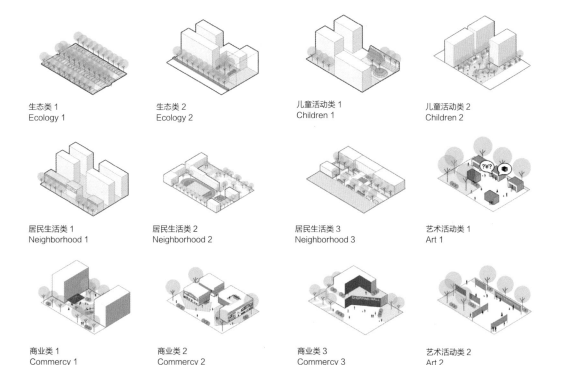

生态类 1 Ecology 1　　生态类 2 Ecology 2　　儿童活动类 1 Children 1　　儿童活动类 2 Children 2

居民生活类 1 Neighborhood 1　　居民生活类 2 Neighborhood 2　　居民生活类 3 Neighborhood 3　　艺术活动类 1 Art 1

商业类 1 Commercy 1　　商业类 2 Commercy 2　　商业类 3 Commercy 3　　艺术活动类 2 Art 2

节点处理模式
Node Design Mode

规划设计 PLANNING & DESIGN

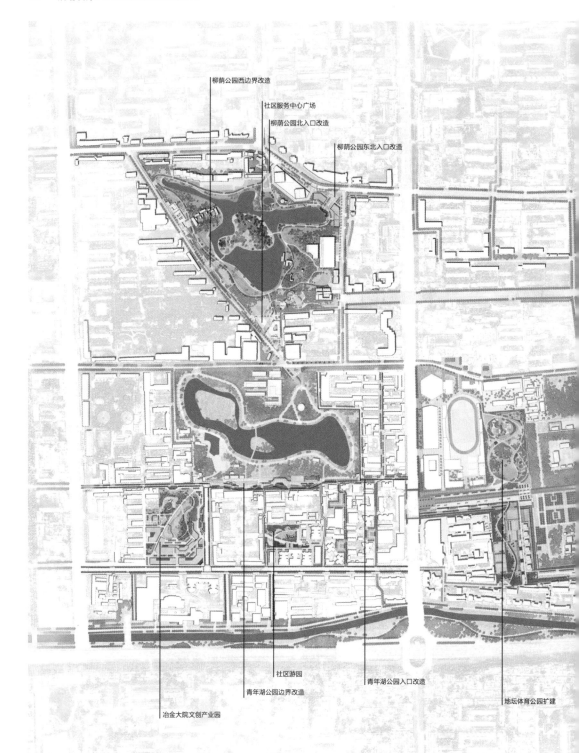

规划总平面
Master Plan

规划设计 PLANNING & DESIGN

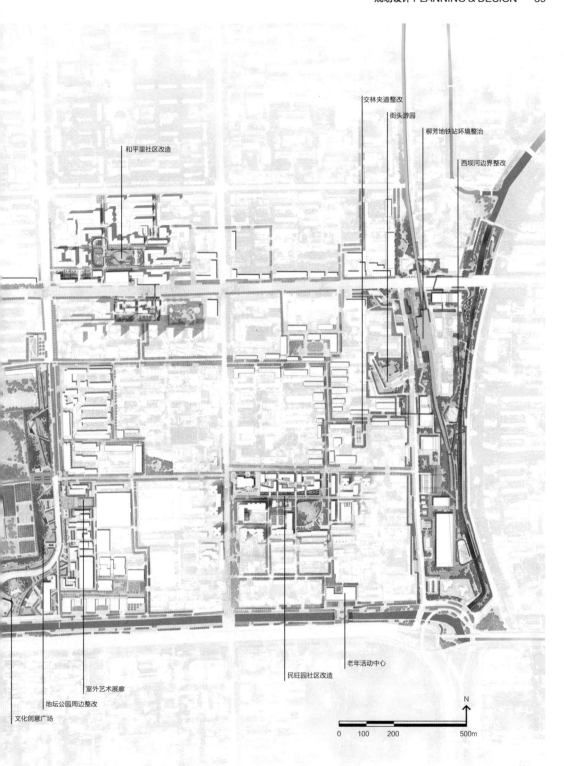

1. 柳荫公园边界改造

柳荫公园周边以居住为主，本次设计主要解决公园连接和公园封闭两大问题：柳荫公园是奥体中心与北二环绿道连通的重要节点，但内部道路曲折；此外，公园现状被围墙封闭，仅有三个出入口。

公园北侧界面：拆除低矮老旧建筑，微调围墙以拓展入口空间。通过空间的串联，将其作为具有社区氛围的公园次入口。同时拆除东北角低矮商业棚户，新增商业建筑，布置室外餐吧，营造亲水空间，将其作为极具商业氛围的公园主入口。

柳荫公园与青年湖公园衔接处：将棚户区改造为社区中心广场，同时改造街角空间，将青年湖公园入口西移，以更好地连接两个公园。在临街一侧布置多功能广场、社区服务中心等营造开敞空间，其中服务中心兼具图书馆、展览、服务问询和餐饮功能，同时注重室外空间的营造。在临近居住区和公园一侧布置运动场、儿童游戏场等林下空间。

公园西侧界面：该区域处于历史文化选线上，北部结合现状餐饮承接老北京美食街的商业氛围，南部承接社区广场生活氛围。设计将柳荫公园柳文化与生态文化在边界进行整合，集商业、科普、文创、售卖为一体，营造活跃的公园边界。

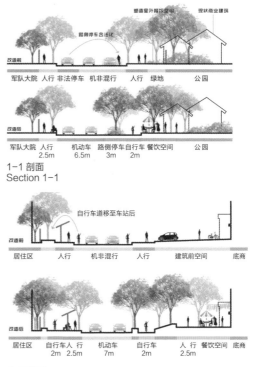

1-1 剖面
Section 1-1

2-2 剖面
Section 2-2

柳荫公园边界改造设计平面
Design Plan of Liuyin Park

规划设计 PLANNING & DESIGN   61

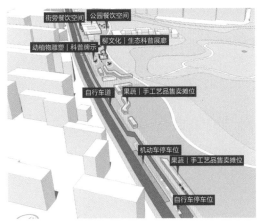

设施分析
Facility Design Analysis

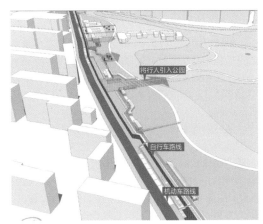

流线分析
Transportation Design Analysis

北入口改造效果
Perspective of North Entrance

新增东北入口效果
Perspective of Northeast Entrance

旱喷效果
Perspective of Water Jets

集市效果
Perspective of Street Market

展廊效果
Perspective of Exhibitions

文创摊位效果
Perspective of Market Stalls

2. 青年湖南街区改造

区域位于城市北护城河北侧，总规中定位为公园游憩区。承接总规"绿地＋"规划理念，依据花园式社区和街区式社区理论，提出将街区内的公园游憩与居住、商业办公等功能嵌套叠加，打造为花园式街区。

通过实地调研，发现区域内的现状冶金社区最具改造提升潜力。冶金社区建筑年久失修，已拆迁完毕，相关资料证实，政府已决定将其全部拆迁，因此社区所有建筑均可拆除，拆除后可改造面积 $5.2hm^2$。用地南边为洲际大厦等高端商业办公建筑，另外三面相邻居住区，周边文教机构丰富。因此不同于以往居住区翻新策略，设计将原有居住用地重新划分为商业办公用地和公园绿地，叠加周边使用功能。根据城市功能性质在南边新增商业、办公综合体和社区中心，同时屋顶绿化全覆盖，满足周边人群使用需求。

通过花园式街区设计贯彻"绿地＋"的规划理念，增加屋顶绿化等绿化面积、充分发挥绿地的生态效益和社会效益。通过合理改变用地性质增加政府税收，提升场地及周边经济价值。通过雨水收集和径流控制，完善城市 LID 系统。

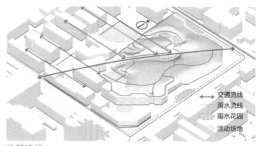

功能流线
Space-streamline Analysis

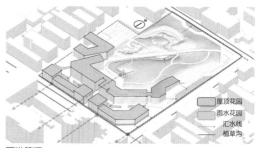

雨洪管理
LID System

儿童活动园效果
Perspective of Children Playground

花园商业街效果
Perspective of Garden-business Street

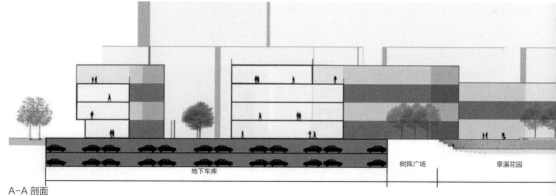

A-A 剖面
Section A-A

规划设计 PLANNING & DESIGN

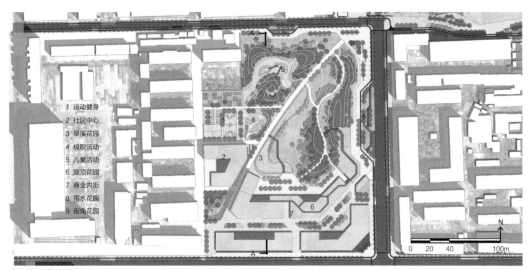

1 运动健身
2 社区中心
3 旱溪花园
4 极限运动
5 儿童活动
6 屋顶花园
7 商业内街
8 雨水花园
9 街角花园

青年湖南街改造平面
Design Plan of Qingnianhu South Street

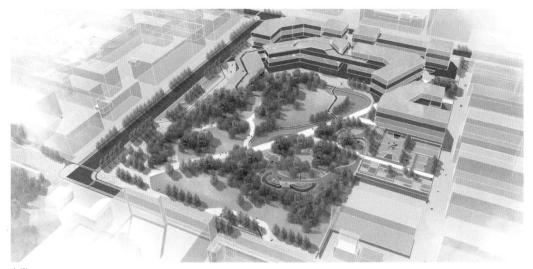

鸟瞰
Aerial View

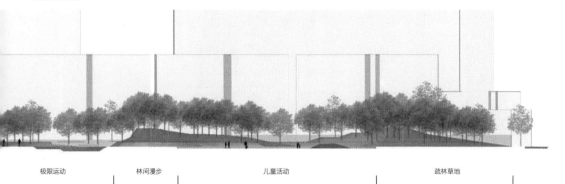

| 极限运动 | 林间漫步 | 儿童活动 | 疏林草地 |

3. 地坛体育公园扩建

地坛体育馆与社区公园设计位于片区中心位置，整个改造范围为 9.8hm²。

周边建筑主要为居住区和学校。设计地块内建筑功能复杂，西侧以商业办公建筑为主；中部以体育功能建筑为主；东部是地坛医院搬迁后的废弃地。地块紧邻地坛公园和青年湖公园，因此承担着联络三大公园的功能。此外，地块周边交通便利，步行通畅。主要设计地块三面围合，进入以步行为主。使用人群基本为附近居民和学生。地块内的地坛体育馆及周边建筑主要承载了东城区专业赛事举办的功能，同时为周边居民提供了室内常规运动的场地。

因此结合上位规划，地块设计应以非常规运动为主，更加侧重于儿童和青少年的运动，与现有功能互补。设计之后的地块将满足周边全年龄段的居民特别是儿童和青少年的户外体育锻炼活动。

设计策略是结合地形设计一条立体的环形跑道，增加跑道长度和趣味性。同时地形还分割出功能分区，包括青少年运动区、极限运动区、儿童游戏区、公共活动区和休闲活动区。

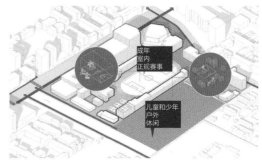

设计目标
Objective

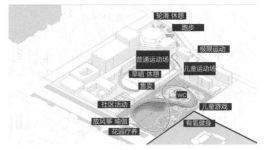

活动类型
Activity Type

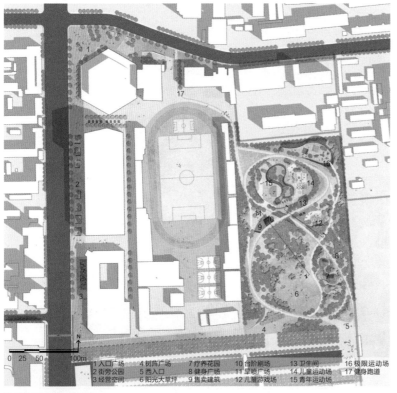

地坛体育公园设计平面
Design Plan of Ditan Sport Park

设计策略
Strategy

规划设计 PLANNING & DESIGN

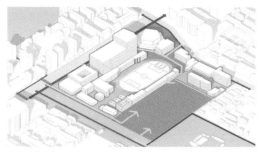

周边交通分析
Traffic Analysis

街道改造效果
Perspective of Street Update

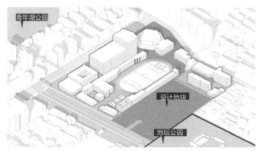

周边绿地分析
Green Space Analysis

公共广场效果
Perspective of Square

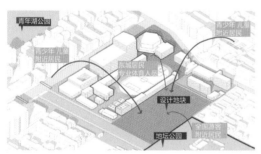

周边人群分析
Population Type Analysis

极限运动场效果
Perspective of Playing Field

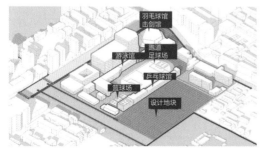

周边功能分析
Architectural Function Analysis

儿童运动场效果
Perspective of Children Playing Field

4. 地坛公园周边整改

地坛公园周边整改延续了上位规划中"绿地+文化"的设计定位。地坛围墙内部为历史遗迹保护范围,但考虑在大型节事活动时适当开放。围墙外围为规划绿地,是本次设计的重点区域。

设计时充分利用现状周边的历史资源和文化艺术景点,并且注意构建与周边大型绿色组团的连接。最终,在地坛公园周边设计一条活跃的景观带。另外,对周边交通及停车状况进行分析发现,日常情况下的停车需求基本满足。但在地坛举行大型节事活动时,则呈现停车位不足、交通拥堵的状况。

针对本地块在日常性活动和事件性活动下动态变化的场地使用状况,提出以活动与事件为主导的弹性场地体系设计方法。

策略一是进行活动事件的策划,分为日常性活动和事件性活动。策略二是弹性场地体系的设计。开放型场地上可以放置临时构筑物或可移动装置,配合各种类型的室外活动;置换型场地仅在大型节事活动时作为停车场使用,日常则作为市民活动的场所。两种策略通过分层叠加的方式得到设计后的活跃景观带,也分为了日常型和节事型。

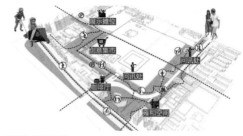

日常性活动分析
Daily Activity Analysis

节事性活动分析
Festival Activity Analysis

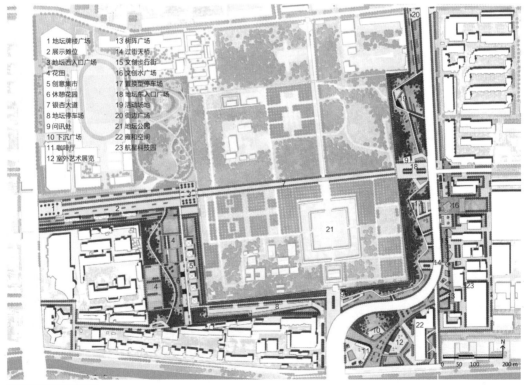

1 地坛牌楼广场　　13 树阵广场
2 展示摊位　　　　14 过街天桥
3 地坛西入口广场　15 文创步行街
4 花田　　　　　　16 文创水广场
5 创意集市　　　　17 置换型停车场
6 休憩花园　　　　18 地坛东入口广场
7 银杏大道　　　　19 活动场地
8 地坛停车场　　　20 街边广场
9 问讯处　　　　　21 地坛公园
10 下沉广场　　　　22 雍和空间
11 咖啡厅　　　　　23 航星科技园
12 室外艺术展览

地坛公园周边设计平面
Design Plan of Ditan Park

规划设计 PLANNING & DESIGN 67

弹性场地策略
Elastic Ground Strategy

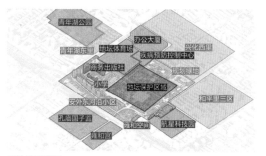

用地性质分析
Land Use Analysis

历史文化资源分析
Historical and Cultural Resources Analysis

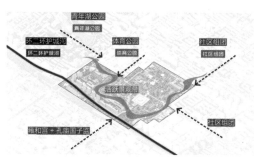

连接组团分析
Connection Group Analysis

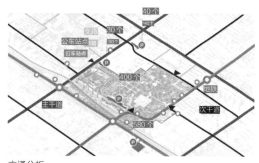

交通分析
Traffic Analysis

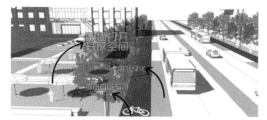

街道空间改造效果
Perspective of Street Space Update

创意集市效果
Perspective of Creative Market

文创街效果
Perspective of Cultural and Creative Street

艺术展览效果
Perspective of Art Exhibition

5. 和平里社区改造

地块北部以居住区为主,包含兴化西里、化工大院、和平里一区、七区等社区,社区中多为20世纪七八十年代新建或翻修的老建筑。在改造中考虑选择适合打开的社区进行改造,增强城市绿网连接度。

和平里七区有小学、宾馆、事业单位等较多公建,本身就已经较为开放,社区中人流量最大的几条流线也是因为这些公共建筑的存在而形成的。设计分析逻辑如下:①设计时依据最大人流量的流线,划定了大致的重点改造区域,对区域中的老旧建筑及棚户进行拆迁;②选择出被原有建筑阻隔、连接东西或南北交通的最短路径,用以构成路网的基本骨架,方便社区与外部城市道路的连接;③在构成的路网骨架基础上增设一条散步道,将被主路切碎的各个功能区块联系起来;④以绿岛、树池等形式保留改造区域中长势良好的植物。

和平里一区在定位上将其改造成类似太古里的商业综合体。设计分析现状流线、拆除老旧建筑,将现状的天桥作为连接的依托,丰富了这条游线(平面图中的红色道路)的竖向景观:先经过两个屋顶花园,再沿建筑的外置阶梯进入下沉花园,最后爬升到抬高的地形观景台,营造出丰富的空间体验。

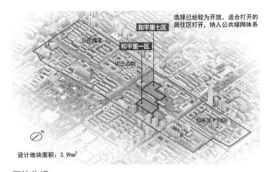

区位分析
Site Analysis

现状分析
Current Situation Analysis

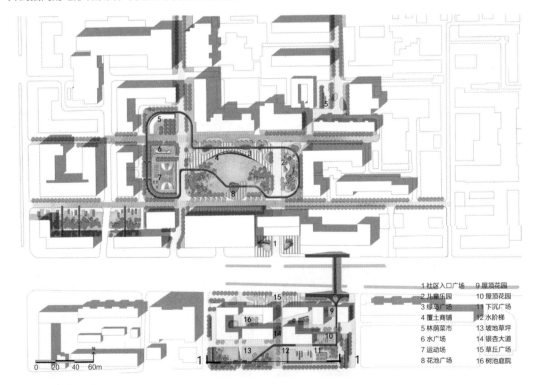

和平里社区改造平面
Design Plan of Hepingli Community Update

规划设计 PLANNING & DESIGN

社区连接分析
Design Analysis

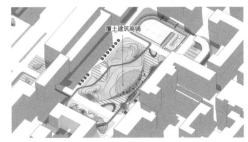

覆土建筑分析
Design Analysis

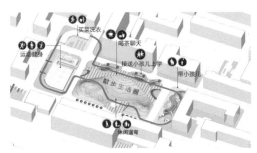

散步生活圈分析
Design Analysis

商业广场设计分析
Design Analysis

中心广场效果
Perspective of Square

下沉花园效果
Perspective of Sinking Garden

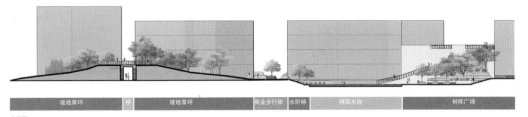

剖面
Section 1-1

6. 民旺园社区开放空间改造

上位规划作为新增服务整个居住片区的绿色核心，占地面积 4.5hm²。意图在对场地进行整改的基础上最大化地发挥其潜在价值，激发社区活力。

设计从现状分析入手，承接上位规划对现状建筑、流线、潜在功能进行分析，结合建成目标，确定整改策略，最终完成设计。现状场地分为三大部分：棚户区、民旺园社区和街角商业建筑，根据目标导向对三块用地提出整改定位：棚户区和民旺园社区建筑年代久、质量差，保留价值不大，因此全部拆除，临街新建商业办公建筑；街角公共建筑可保留作为社区服务中心。

对周边建筑出入口和流线进行分析后发现，人车混行问题严重且多处道路不通畅，在设计中将人车进行梳理分流，车行从场地外部进入小区，内部仅供自行车和人行使用。此外，在考虑地上停车的基础上设置了地下停车以缓解停车压力。

结合功能需求将场地分为四大区块：商务办公开放空间、居民休闲区、运动健身区和社区中心，为场地赋予多种复合功能。中部步道主轴串联了四个片区，形成了一轴四区的格局。

细部设计中，北部商业办公建筑形成多条内街和中心庭院空间，中部的旱溪花园可以收集建筑屋顶雨水和主园路雨水，经湿生植物净化后收集利用。

基于"绿地+"的概念，通过开放社区、激活商业、融合生态的手段，创造功能复合的新型社区开放空间。

区位分析
Site Analysis

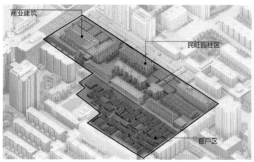

场地分析
Site Analysis

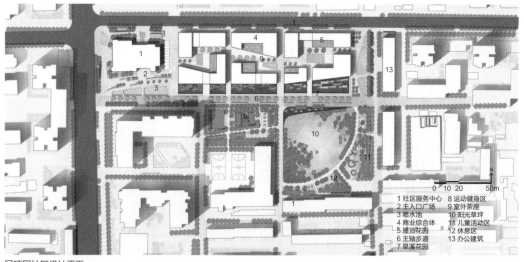

民旺园社区设计平面
Design Plan of Mingwangyuan Community

规划设计 PLANNING & DESIGN

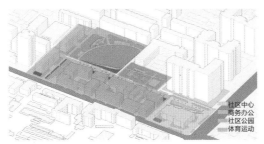

功能分析
Function Analysis

旱溪花园效果
Perspective of Dry Creek Garden

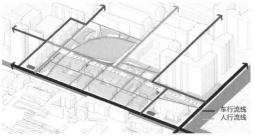

交通分析
Traffic Analysis

建筑中庭效果
Perspective of Building Atrium

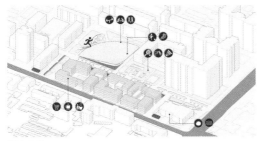

活动分析
Activity Analysis

主入口广场效果
Perspective of Plaza

鸟瞰
Aerial View

7. 西坝河片区开放空间改造

西坝河区域的主要问题包括：驳岸硬质化，缺乏亲水性；河道与城市界隔离，联系性较弱；滨河空间利用率低，活动场地较少；居住和商业建筑占地多，活动类型少。根据现场问题，提出结构连通、多类型袖珍公园与生态型河流绿地相结合的改造策略。

现状场地中包含了餐饮、居住、医疗、办公、零售、停车等多种功能，周边有快速路、主干道、次干道、支路和居住区道路，通过对小区的出入口分析，呼应场地的通行设计，分析场地性质，将地块分为生态自然区、生活休闲区、滨水游憩区和商业活力区。东西向街道现状问题较多，机动车占用了场地大量的空间，街道拥堵且缺乏活动空间。通过规划地上与地下停车场、增加自行车道、打破围墙的方式来解决现状问题，从而打开城市与西坝河的连接，梳理南北向通行。

场地南部拆除了四个现状较差的建筑，形成三个东西向连接的潜力地块，根据现状情况规划活动类型，在潜力地块一和三处，以桥梁和高架的形式连接西坝河两岸，潜力地块二使城市界面能通往西坝河，三处潜力地块共同形成了南部场地的连接结构，高架连接西坝河两岸，并与现状改造建筑形成的屋顶花园相接，桥梁连接回旋坡道通向滨河区域，连通西坝河两岸，同时规划多处停车场以及自行车场来梳理南北向通行状况。

北部拆除五个现状较差的建筑，形成两个东西向连接的潜力地块，由于东部高档社区的阻隔，只能形成东西向的连接，并且改造现状天桥，增加绿色基础设施，形成南北地块的主要连通区域。同时，规划多处停车场，并且在社区活动广场设置地下停车场。此外，整个场地根据功能分区种植不同类型的植物。

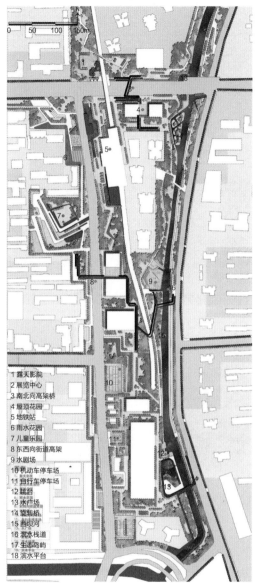

西坝河片区平面
Design Plan of XiBa River Area

入口分析
Entrance Analysis

分区分析
Subarea Analysis

功能分析
Function Analysis

道路分析
Road Analysis

规划设计 PLANNING & DESIGN　73

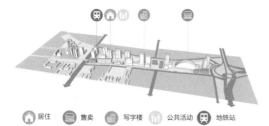

现状建筑功能分析
Architectural Function Analysis

河道剖面
Riverway Section

南部功能对比分析
Southern Functional Comparison Analysis

儿童游乐效果
Perspective of Children Playground

北部功能对比分析
Northern Functional Comparison Analysis

露天观影效果
Perspective of Outdoor Cinema

跨河交通分析
Across River Connection Analysis

鸟瞰
Aerial View

The site is located in the northern part of Dongcheng District, and is separated from the inner city by a moat. It connects Andingmen and Lama Temple in the south, and adjacent to Drum Tower Street in the east. The aim of this project lies on the topic how to analyze and transform the site and strengthen the characteristics of the site under the background of stock planning policy and urban double-repair, with the view of "constructing green ecological network".

The construction of the site began in the 1960s, and the land is dominated by residential areas. The facilities are complete and have the characteristics of closed, and complex. The site is generally flat and the terrain is lower than the outside. Under the background of frequent extreme climates, the stress of rainfall and pollution increases. On the socio-economic level, the aging society tends to strengthen; the pressure of population flow is high, and the corresponding infrastructure is insufficient. There are many green space resources in the site but the distribution is uneven. The green space on the west side is relatively large, while the green volume on the east side is insufficient. The overall green space connectivity is insufficient, and the overall environmental quality needs to be improved.

However, the public transportation around the site is convenient, and tourism resources are rich in historic sites which date from the Ming and Qing Dynasties. Many architectural complexes constructed in the "planned economy" era are in good condition. Therefore, how to construct the base of green infrastructure and reflect the historical and cultural characteristics of the site is the focus of our design under the government policy of Dual-repairing and Open Communities.

Through qualitative and quantitative overlay screening, planning and design will coordinate ecological, historical, and transportation requirements, and a green slow traffic, functional compound open space network system will be created.

By the five steps of reviewing fractured green space, tapping potential space, constructing an ecological base network, establishing theme path, and deepening the theme space, we will gradually realize the construction of the "Green Gap Space Infiltration System". Applying the visual parameter design platform, the reforming scope of the ecological network base will be determined. On this basis, the three major theme paths of the green open space system are defined, including an ecological network, a multi-population adapted greenway, and a memory-carry historic path. These three Superimposed levels form the specific scope, strategy and content of planning and design.

Based on the comprehensive analysis of the previous period, it is determined that the Ditan and the Lama Temple are the two major attractive centers in the area. Through the combing of historical elements and elements of natural water resources, the historical connection zone of "Waiguan Street-Ditan Area" and the "Two Belts" of the " City Moat -Xiaba River" urban waterfront scenery belt are constructed, combining with the three main north-south direction roads to repair the site.

Under the overall planning structure of "Two Hearts, Two Belts, Three Verticals, and Multiple Nodes", the three major themes and the seven regions' distinctive green features are deepened, so as to achieve the green penetration in the dense urban area.

共享绿色的实现
渗绿小空间体系构建
Sharing Green
Green Space Penetration System Construction

贾子玉、王念、巴雅尔、周超、孙悦昕、卓荻雅、霍曼菲
Jia Ziyu / Wang Nian / Ba Yaer / Zhou Chao / Sun Yuexin / Zhuo Diya / Huo Manfei

北中轴在北京城市规划中具有"城市地标"的特殊意义，其周边用地的发展影响着北京城市特色的体现。

场地位于北京市东城区北部，与内环古城相隔一条护城河。场地内既有柳荫公园、青年湖公园等开放活动空间，也有地坛等重要历史文化节点，但由于交通割裂、用地局限等原因，场地内部景观状况参差不齐，现状绿地早已无法满足居民日益增长的休闲文化需求。

本次规划设计的思考重点在于如何在存量规划和城市双修的背景下，以"绿色生态网络构建"的视角，对场地进行分析和改造，强化场所特征。以"缝中求绿，渗透延续"为主题，打造符合生态修复、历史文化保护、城市游憩需求的绿色慢行开放空间系统。

场地位于北京市东城区北部，与内环古城相隔一条护城河。南接安定门和雍和宫，东临城市中轴线——鼓楼大街。本次项目将在存量规划和城市双修的背景下，利用"绿色生态网络"对场地进行分析和改造。同时，强化场所特征，构建绿色基础设施基底，体现场所历史文化特色。

场地整体平坦，地势较低，雨污胁迫大，于20世纪60年代开始兴建，用地以居住区为主，多是单位大院，配套设施齐全，具有封闭性、复杂性、复合性的特点。从社会层面来看，老龄化趋势加强，人口流动量大，相应基础设施建设量不足。就现状户外空间品质而言，场地周边公共交通便利，绿地资源较多但分布不均，西侧公园绿地服务覆盖度大，而东侧绿量不足，绿地间连接度不够，总体环境质量需要提升。另外，场地古迹旅游资源丰富，诸多计划经济时代建成的单位大院建筑风貌完好。

规划设计通过定性、定量的叠加筛选，协调绿色、历史、交通三方面的要求，构建功能复合的开放空间网络体系。

通过审视破碎绿地、挖掘潜力空间、构建生态基底网络、建立主题游线和深化主题空间五个步骤，逐步实现"绿色缝隙空间渗透体系"的构建。应用可视化参数设计平台，确定生态网络基底的改造范围。在此基础上，确定绿色慢行开放空间体系的三大主题游线——便捷生活的生态网络基底、多人群适应的功能绿道和承载记忆的历史游线。这三个层次的叠加形成了规划设计的具体范围、策略与内容。

规划前，通过大量的实地调研，确定可改造的建筑及户外空间，叠加现状绿地及可改造的户外空间，得到改造后所有绿色空间并为其分级，这些绿地成为绿道构建中潜力较大、可考虑连接成线的"点"；同时，发掘所有开放区、社区单位内部的人行路线，使这些路线成为具有改造潜能的"线"。

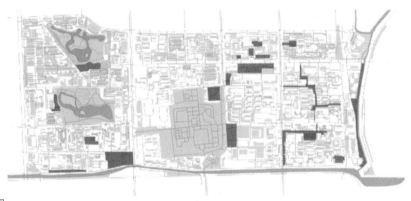

主要改造空间

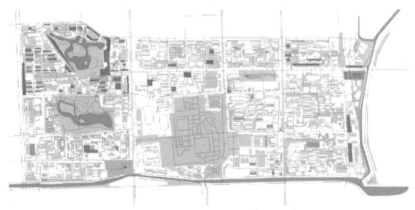

次要改造空间

户外环境改造潜力点分析
Potential Rebuilding Areas Analysis

在绿道选线方面，便捷生活圈选线为居民提供到达交通站、学校、公园、商业等生活必需地点的快捷通道。在此过程中，通过软件计算各个节点之间的最短路径，筛选出最短路径网络，叠加路径所串联的区域内各类资源，得到便捷生活通勤圈的全部选线。

多人群使用绿道选线主要考虑到改造后形成的公园绿地、街旁绿地、社区附属绿地、带状绿地等绿色空间，旨在为区域内居民提供环境质量更高的功能型绿道，为此区域的游客提供更优质的活动游线，提升区域品质。

承载记忆的历史游线选线考虑到区域内重要的历史节点，如外馆斜街、雍和宫以及质量较好、具有年代感的社区建筑，将这些历史元素以及周边地铁站串联成为历史游线绿道，提升环境配套设施，为区域旅游发展提供保障。

人行空间分析
Walkable Space Analysis

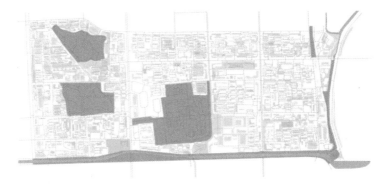

改造后主要绿地

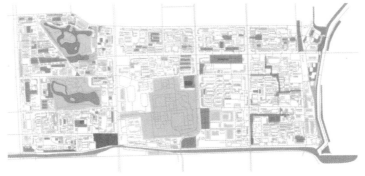

改造后次要绿地
绿色空间改造
Green Space Transformation

便捷生活通勤圈选线
Shortest Walk Path Analysis

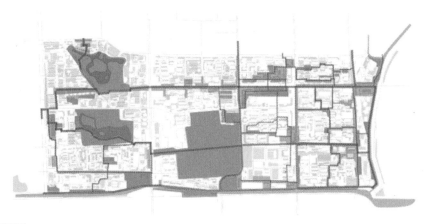

多人群使用绿道选线
Multi-use Path Analysis

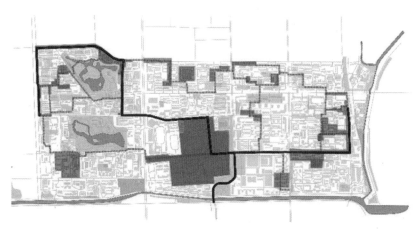

历史游线选线
Historical Path Analysis

规划设计 PLANNING & DESIGN

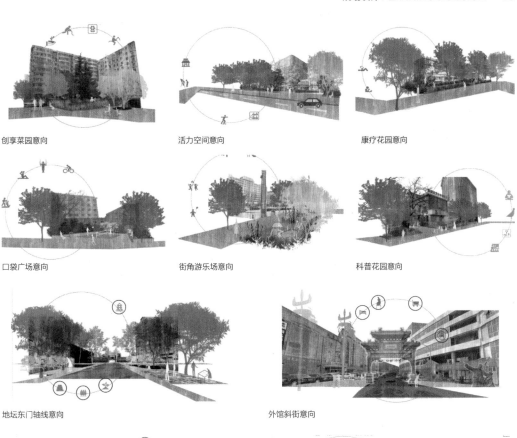

创享菜园意向　　　　　　　活力空间意向　　　　　　　康疗花园意向

口袋广场意向　　　　　　　街角游乐场意向　　　　　　科普花园意向

地坛东门轴线意向

外馆斜街意向

5个老小区意向
改造单元模式
Unit Reform Mode

城市发展穿越线意向

　　最后，依据三类绿道的不同要求，在其中选取可改造的潜力点，提出设计策略。便捷生活圈作为绿色生态网络基底，需要满足整洁干净、绿量充足等最基础的设计要求。对杂乱的街道立面以及使用效率低、无绿化的停车场进行整改；条件允许的建筑增加屋顶绿化及垂直绿化，提升区域整体绿量。

　　多功能绿道叠加于便捷生活圈之上，居住区周边绿地布置创享菜园供居民管理使用，医院周边绿地布置康疗花园种植芳香、康养型植物，科普花园与街角游乐场可设置在学校及居住区周围，为青少年提供游乐活动、室外学习场所。

　　历史游线绿道同样叠加于便捷生活圈之上，区域内包括地坛公园、外馆斜街等著名历史遗址，也拥有风貌典型、年代感强的老社区，本绿道主要为区域内的重要历史节点服务，同时通过历史游线的串联，为游客展现计划经济时代建成的社区风貌。

　　另外，于地坛东门增加街头绿地广场，更换与地坛相呼应的地面铺装，两侧种植大乔引导视线；外馆斜街进行建筑材质统一、店招统一、蒙古元素雕塑牌楼设立等改造；5个老小区中注意保留建筑风貌，维修复原，增加宅间绿地开放性与功能性；绿道中间段恰可穿越1950年至今的老中新三类小区，增加导览、雕塑等景观构筑，使其连续为一条城市发展穿越线。

规划设计 PLANNING & DESIGN

规划总平面
Master Plan

规划设计 PLANNING & DESIGN

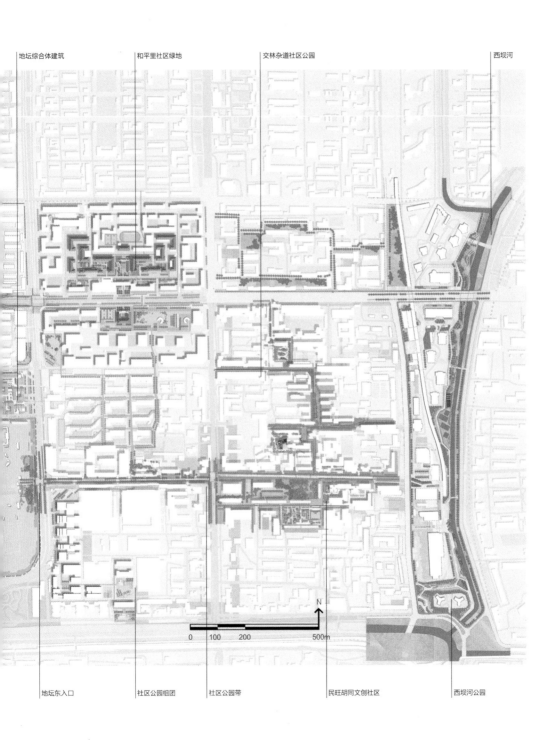

1. 历史慢行体系设计

地块南部为奥林匹克森林公园及北土城带状绿地，北部毗邻护城河公园与古城区。规划中将该地区定位为历史游览主题的慢行系统，串联外馆斜街、总政大院、安德里社区绿地以及冶金大院三个历史节点及一个绿道节点。

针对外馆斜街节点，通过拆除阻碍性及低品质建筑、打通柳荫公园与安华路带状公园绿地形成了贯穿南北城市绿道及视觉通廊，并给绿地赋予商业、游憩、历史地标等多重功能。

在黄寺总政大院内中心绿地中，包含有公共服务建筑一处。设计在公共建筑内附加副食店的功能，打造社区服务中心，重现历史记忆中的"大院"。另外，将相邻两块小型场地通过道路引导整合，引入儿童游乐、运动健身、水景游乐等活动，增加绿地活力。

安德里社区绿地现状功能单一，周围以居住建筑为主。将东部现状商业建筑拆除，新建开放式社区集市，为流动商贩和社区居民提供商业场地，打造沿街活力界面。社区内还布置创想菜园、社区剧场、社区纪念园等重要节点，为居民生活提供便利。

1 公园入口
2 茶室
3 草坡
4 滨水绿廊
5 街旁绿带

外馆斜街改造平面
Design Plan of Waiguan Street

公园北入口效果
Perspective of North Extrance

柳荫公园东入口效果
Perspective of East Entrance

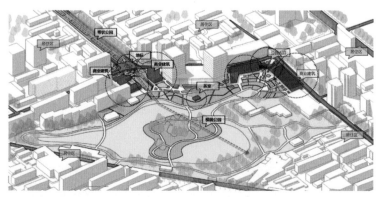

外馆斜街鸟瞰图
Aerial View of Waiguanxiejie

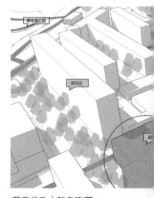

黄寺总政大院鸟瞰图
Aerial View of Huangsi Commun

规划设计 PLANNING & DESIGN

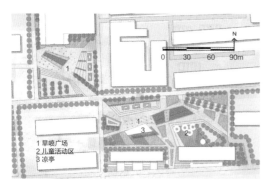

黄寺总政大院改造平面
Desgin Plan of Huangsi Community

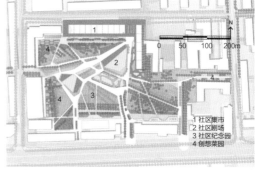

安德里社区绿地改造平面
Design Plan of Andeli Square

黄寺总政大院中心绿地效果
Perspective of Huangsi Community

安德里社区雨水花园效果
Perspective of Andeli Community

黄寺总政大院儿童游戏区效果
Perspective of Children Activites Area

安德里社区创想菜园效果
Perspective of Vegetable Garden

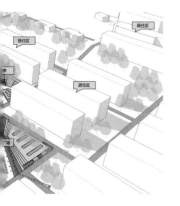

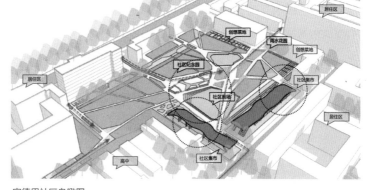

安德里社区鸟瞰图
Aerial View of Andeli Community

2. 青年湖公园周边改造

安德里北街及周边街区有三个城市综合性公园，紧邻护城河，是周边居民与外地游客日常活动与游览的主要片区。基于上位规划，为了强化沟通公园与周围住区的连通性，考虑在各个公园出入口增加景观廊道，构建绿色通道，串联各街区的公共活动空间。绿色通道是串联各个绿色基础设施的主要道路，主要由三类道路构成，分别是连通绿道、渗透绿轴和公园边界散步道。在绿色通道的基础上，结合各个社区的外排水，形成不同等级的雨洪通道，达到给公园补水的作用。核心开放街区与公园及绿色通道相连，为周边居民提供日常休闲活动的便捷场所，通过改造片区景观，提高街区生活环境品质。本片区集中在公园周边设置了三个重点改造地段，设置了多个开放空间，满足不同的人群功能需求。例如，利用安德路北小区内原有的社区绿地打通柳荫公园南侧的东西向景观廊道，拆除部分棚户，同时设置沟渠解决场地雨洪问题；历史上柳荫公园因烧制陶土挖出的蓄水坑，因此该地段主要以旱溪浅滩呼应历史水系，串联大小坑塘，构成连续的景观空间。

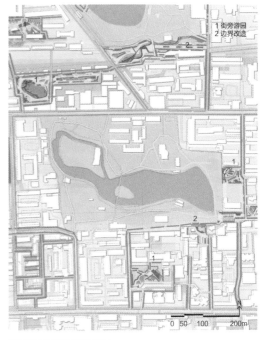

青年湖公园周边设计平面
Design Plan of Qingnianhu Park

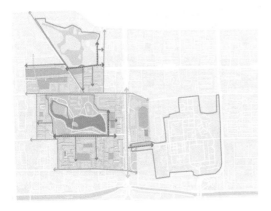

绿色通道分析
Greenway Analysis

雨洪管理分析
Stormwater Management Analysis

旱溪街区效果
Perspective of Block

叠水广场效果
Perspective of Waterscape

规划设计 PLANNING & DESIGN   85

拆除棚户区，迁移公园管理处

打造生态空间

设置旱溪蓄积住区雨水

打造特色集市区

拆解部分建筑，腾留活动空间

合理安排停车空间

流动性商铺功能集中化

提高公园入口景观辨识度

青年湖公园北侧设计策略
Design Strategy of North of the Park

青年湖公园南侧设计策略
Design Strategy of South of the Park

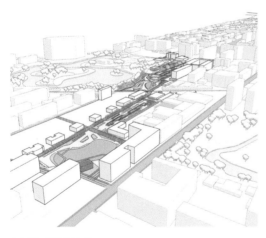

青年湖公园北侧鸟瞰
Aerial View of North of the Park

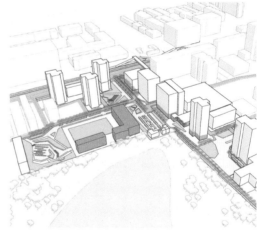

青年湖公园南侧鸟瞰
Aerial View of South of the Park

集散广场效果
Perspective of Square

科普花园效果
Perspective of Garden

3. 和平里北街周边改造

和平里北大街位于北京的中心地区——东城区。基于街区现状分析，可以得出，目前该地区的交通系统相对完善，并拥有一定的水体及绿色空间。但是，街区内依旧存在不少共性问题，需要通过更好的城市绿地规划来改善。例如，街区内缺乏足够的停车位，车辆乱停放问题严重；人行道安全性低，慢行舒适度较差；居民生活圈缺乏绿色空间，且现状绿地的公众利用率较低等等。

为了解决上述问题，规划中提出了四点策略：①打通城市内部微循环，实现道路畅通，建立舒适的慢行体系，加强步行体系间的联系，使步行和骑行成为大多数人的出行选择；②增强绿地环境的使用效率，加强绿地之间的联系，使环境的地域性特色得到体现；③提供丰富的活动场所，丰富人们的社交体验，打造充满活力的公共空间；④重视文化遗产及具有历史意义的区块，使不同年龄和社会地位的人们能够共享安全与活力。

街道是兼具共享与交互的公共场所，人们所感知到的街道空间，应该是安全、友好、舒适且利用率高的空间。绿地应当提供积极、健康的环境，促进人们的身心健康发展，促使社区成为一个充满关爱和包容、富有创意与活力、拥有丰富文化体验的空间。城市环境也将因为这些绿色服务设施的存在，成为一个更具有互动体验和人文活力的共享家园。

1-1 剖面
Section 1-1

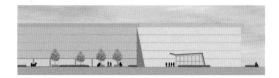

2-2 剖面
Section 2-2

3-3 剖面
Section 3-3

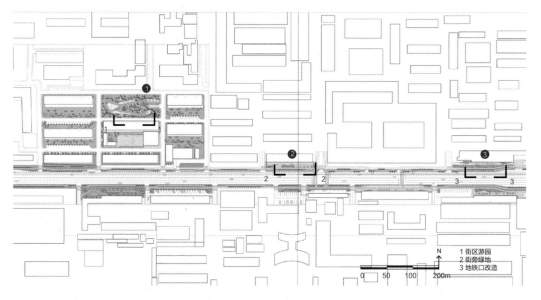

和平里北街设计平面
Desgin Plan of Hepingli North Street

规划设计 PLANNING & DESIGN

小公园内部效果
Perspective of the Park

小公园西侧外部效果
Aerial View of the Park from West Side

小公园外部效果
Aerial View of the Park

和平里北街地铁站效果
Perspective of HePingLi North Street Subway Station

残疾人停车场效果
Perspective of Parking Lot for the Disabled

4. 综合体设计

地块位于地坛东北部的一块废置绿地内，周边有零散的超市与菜市场。地块分为地上、地下两部分，地上部分的北街有一栋商住结合建筑，业态较为杂乱，中间为一块废置绿地，西面与南面多为办公场所，绿地荒废，且功能单一；地下部分现存一个早期防空洞，采光条件和消防条件较差，存在着一定的安全隐患，现作为早市使用。

综上分析，建设良好的绿化、办公与商业环境是场地的迫切需求，故结合周边现状进行小型综合体设计，针对现状问题提出改造建议：①防空洞改造中，考虑地下空间的通透性，同时兼顾上下空间的联系性，将阳光引入地下，结合屋顶绿化，兼顾生态和观赏效益；②打破围墙，通过对原有围墙的拆改，让基地内更通透，充分利用围墙本身，使之兼顾实用性和娱乐性；③提升场地的绿化空间品质，不仅在地表一层打造绿色空间，还将种植引入地下一层；④对北边的建筑进行局部改造。将建筑底层打开，形成良好的交通空间，以提升场地内部活力。

设计对综合体整体进行拆分。将场地分成四大块，为了更好的采光，在中间采用类似天井的做法，将自然光引入建筑；接着，为了让游客有更好的视觉效果，对每个建筑体块的一角进行降低，并进行体块的具体细化和推敲；随后，把四个体块用连廊连接，加强交通联系的同时也增加观赏外部景观的空间；最后，推敲景观与建筑的关系，使两者充分融合。

效果 1
Perspective 1

效果 2
Perspective 2

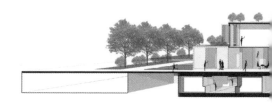
建筑剖面 1
Architecture Section 1

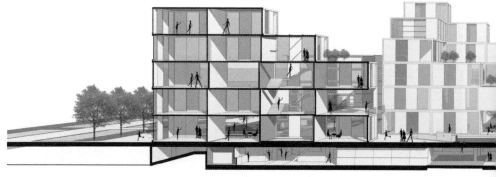
建筑剖面 2
Architecture Section 2

规划设计 PLANNING & DESIGN   89

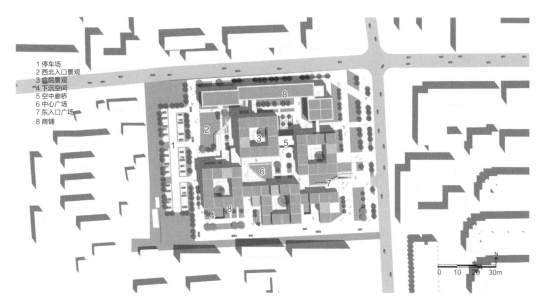

1 停车场
2 西北入口景观
3 庭院景观
4 下沉空间
5 空中廊桥
6 中心广场
7 东入口广场
8 商铺

总平面
Site Plan

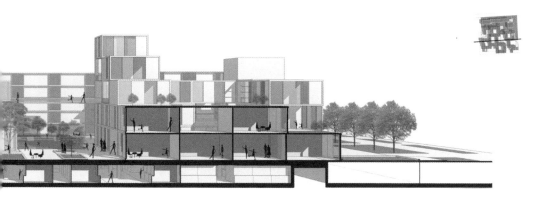

5. 和平里社区步行体系改造

和平里社区是新中国成立初期兴建的现代化住宅区，具有老龄化特征。用地功能复杂，包含很多部委大院等单位。自20世纪70~80年代起，虽一直经历改建更新，但依然保存着片区居民的历史记忆。地块南片是现代化的办公区域，风貌由北向南、从历史向现代过渡。针对场地现状，提出以下设计要点：①增强慢行系统南北向的连接度；②通过穿越各片区的慢行系统，体验不同年代的城市风貌特色，探寻城市记忆。

整个方案由完整的慢行系统串联起四个设计节点：老龄社区、社区游园、和平里一小广场、航星园休闲景观。

和平里七区作为20世纪50年代第一批"现代化住区"，体现了新中国成立初期的城市居住风貌。大街坊、轴线对称、秩序感的空间特点以及结构外露、红砖墙、坡屋顶的建筑特点，无不体现街区的历史记忆，具有重要保护意义。设计主要从梳理交通、更新功能、立面改造和景观提升四个方面进行，对宅间绿地进行重新设计，使之适应老年人交往、健身等需求。

社区游园部分拆除危楼、整合现有绿地，形成完整的社区游园，增加了社区的休闲空间。其中设置运动场、棋牌区、阳光草坪等功能区，并建设一座社区博物馆作为轴线的收尾。

和平里一小门口逼仄狭隘，难以适应人流量大的需求。设计中将两栋危楼拆除，整合户外空间，形成可快速通过的集散广场。同时设计了等候区，方便家长接送孩子上下学时交流休息。

航星园办公区整合边角建筑空间，增加休闲餐饮功能，并面向城市开放，弥补园区餐饮服务设施缺乏的问题。同时将原有篮球场进行再设计，为园区内的工作者提供多样化的活动场所。

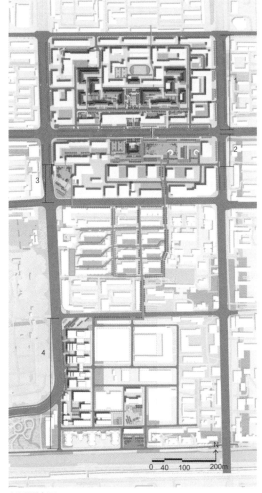

1 和平里七区　　3 和平里一小入口
2 社区游园　　　4 航星园办公区

和平里社区平面
Design Plan of Hepingli Community

和平里社区效果
Perspective of He Pingli Community

社区游园效果
Perspective of Community Garden

规划设计 PLANNING & DESIGN　91

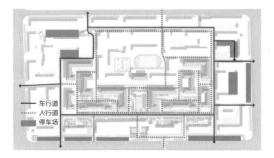

和平里七区交通分析
Heping Li Community Traffic Anaysis

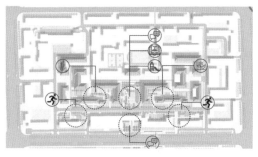

和平里七区功能分析
Heping Li Community Function Analysis

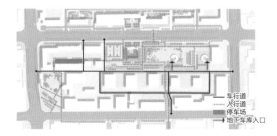

社区游园交通分析
Community Garden Traffic Anaysis

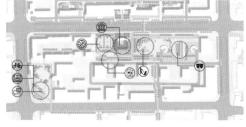

社区游园功能分析
Community Garden Function Analysis

航星园交通分析
Hangxing Garden Traffic Anaysis

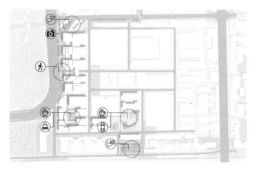

航星园功能分析
Hangxing Garden Function Analysis

和平里一小广场效果
Perspective of the Square of Primary School

航星园效果
Perspective of Hang Xi Yuan Community

6. 社区口袋公园及道路景观改造

该分区用地类型主要为居住用地和教育科研用地。依据建筑和围墙的组合形式，结合规划要求，该分区共包含三个连接性节点、四个公共场所出入口节点以及三个居民活动点的提升改造，设计引入五个口袋公园，形成南北向串联的绿色生活空间，实现"缝中求绿，渗透延续"的概念。

分段一为交林杂道北段入口，东西两侧单位的围墙围合出狭长的线性空间。采用口袋公园的形式作为交林杂道社区居民主要的绿色活动空间，实现"公园－街道－学校"的界面过渡，为使用者提供休息、停留、玩耍的公共活动空间。

分段二东西两侧连接两处社区、一个幼儿园及一个社区大学，两处转折性空间形成改造升级的重点。针对分段整体，进行交通组织，形成"建筑前区－绿色设施带－双向混行机动车道－雨水消纳分车带－自行车道－人行道－附属绿地"的过渡，并对幼儿园内外进行整体空间提升，形成更有趣味、品质更高的儿童活动绿色空间。

为加强整体场地的连通，对设计分段三进行路口渠化改造及道路断面改造。结合现状，引入绿化分车带，使各类交通更有序。

分段四现状为无序管理的棚户区，产业组成为餐饮及装修。据规划要求，该分段需起到串联南北向和东西向绿色空间体系的作用。因此，引入社区公园及文创活动中心，在公共绿地中，通过地形的塑造，增加空间层次，形成活跃的活动场所，文创建筑结合地形，形成丰富的空间体验。

❶ 交林杂道口袋公园
❷ 砖角楼北里露天集市
❸ 东城区卫生局第三幼儿园活动空间
❹ 民旺北社区公园
❺ 民旺北文创胡同

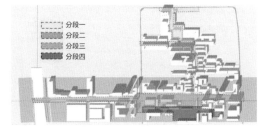

空间分区
Functional Division

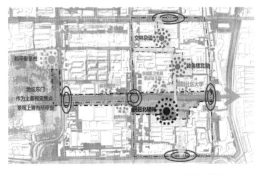

景观轴线及景观节点布局
Landscape Axis and Spots Structure

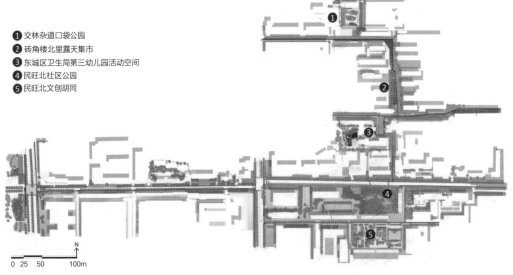

交林杂道－民旺社区周边设计平面
Design Plan of Jiaolin-Minwang Area

规划设计 PLANNING & DESIGN   93

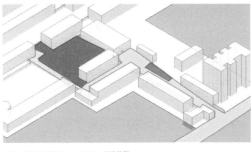

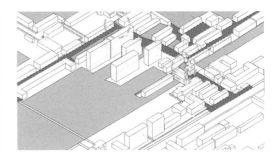

图例:
- 社区卫生服务站
- 现状开放空间
- 教育用地
- 人行道
- 改造范围
- "口袋公园"单元
- "康体花园"单元

分段一现状及改造策略
Section One Analysis and Refrom Strategy

分段二现状及改造策略
Section Two Analysis and Refrom Strategy

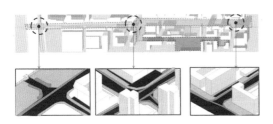

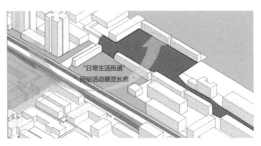

图例:
- 分段三
- 分段四
- 现状人行游憩空间
- 现状机动车空间
- 现状自行车空间

分段三现状及改造策略
Section Three Analysis and Refrom Strategy

分段四现状及改造策略
Section Four Analysis and Refrom Strategy

交林杂道口口袋公园效果
Perspective of Pocket Park

第三幼儿园活动空间效果
Perspective of Kindergarten Activities Space

文创胡同下沉广场效果
Perspective of Sink Square

民旺北社区公园效果
Perspective of Community

7. 西坝河周边改造

场地内以居住及办公建筑为主，混杂部分商业建筑，地块中河道西侧大量建筑违章占用绿地空间，各单位修建的围墙使滨河绿地可达性差、使用率低；机动车流线较为清晰，但人行空间混杂，无法在整体绿地系统中实现连通功能；现状西北侧有条件良好的带状公园，滨河东侧具备较为完整的带状绿地，但西侧绿地未能连成体系，多数为办公区及居住区内部绿地；河道驳岸硬质化严重，水质较差。

基于场地现状，提出三个改造策略：①拆除地块内对人行系统影响较大的商业建筑及滨河绿带中的违章建筑；②将居住区及部分办公区围墙后退，开放部分重要节点的停车场，增加区块内公共空间；③将原有公共空间与新增公共空间串联成体系，增加滨河绿地的可达性及连续性。针对硬质化驳岸，在绿地宽度允许的河段施行缓坡入水改造，局部区段应用生态石笼墙固土护坡，高差过大的南端则保持硬质驳岸，增设亲水台阶、平台。最终，打通东西向多处通道，形成重要的活力地段。

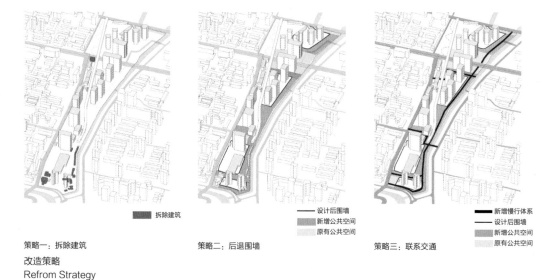

策略一：拆除建筑　　策略二：后退围墙　　策略三：联系交通

改造策略
Refrom Strategy

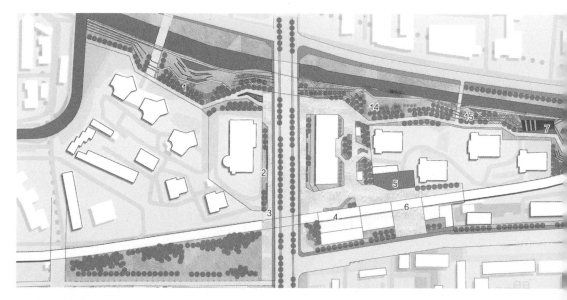

西坝河段设计平面
Design Plan of Xiba River

规划设计 PLANNING & DESIGN

超市前广场效果
Perspective of Square in front of Supermarket

办公区休闲广场效果
Perspective of Spuare in front of Office Area

自然草坡效果
Perspective of Lawn

居住区运动场效果
Perspective of Playground

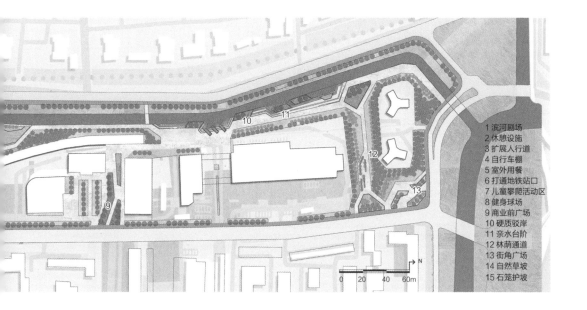

1 滨河剧场
2 休憩设施
3 扩展人行道
4 自行车棚
5 室外用餐
6 打通地铁站口
7 儿童攀爬活动区
8 健身球场
9 商业前广场
10 硬质驳岸
11 亲水台阶
12 林荫通道
13 街角广场
14 自然草坡
15 石笼护坡

The site is located in the east of junction area of Chaoyang, Xicheng and Haidian District, in the west to the West Earth City Road, in the north to the Xitucheng Road, and in the south to Huangsi street and Xin Kang Road. The site is across the north three ring roads and the north central axis of Beijing, with an area about 624 hectares. The north side of this site is the Chinese national garden and the Olympic Sports Center, the west side is the Yuan Ruins Park, the south is the Willow Shade Park, the Youth Lake Park, and the Soil Temple Park. This area is the transition zone of the capital function core area to the urban function expansion area, and bears the responsibility of extending the landscape of Beijing axis and the transition of urban function.

The area was first built since the 1960s. There are Yuan Ruins Park, Ji Men Yan Shu and other historical and cultural nodes. There are also some elements with modern characteristics such as the North shadow studio, the Science and Technology Museum, etc. The site is dominated by residential land., The type of land use is relatively unitary, the distribution of commercial and service land is uneven, and the construction of supporting facilities is relatively inadequate. The number of parks is not enough, which cannot meet residents' daily recreation needs. The distribution of protective green space is uneven. The type of attached green space is mainly residential park. At the same time, the site is full of cultural and historical resources, but there is no corresponding theme space for the nodes such as Jiande Gate and Anzhen Gate. In addition, the commercial the public space of Madian mosque is gradually squeezed. The site is convenient for external transportation, but the horizontal connection between adjacent areas is not good.

As a whole, the site is located along the north central axis and occupies an important location. However, due to the early age of construction, there is a great contradiction between the landscape, traffic, architecture and the living style of the residents, which is urgent to be solved.

Through the study of Beijing greenway system planning and related policies, combining with the analysis of the nature of land use, building types, traffic conditions and historical context, it is concluded that the planning target is "the big connection with the small circulation, city function transition zone ecological green net construction". Through the method of comprehensive stacking and calculation, the potential plots urban greenway will be selected based on spatial syntax, and the final transformation scope will be formed. According to the characteristics of different blocks of the site, the area will be divided into three types: the historical axis area, the theme parks and the living are.

Specifically, The design of the historical axis area is not only to build greenway to solve the current traffic problems, but also a narrative approach to Beijing's central axis landscape transformation upgrade. Three landscape nodes "looking back on history", "washing pond" and "quiet garden", will be designed aiming to create a historical axis from the city to nature. The theme parks include science and technology innovation park and film cultural park. The science and technology innovation park is divided into four main functional areas including core square, science and technology display, potential excitation and scientific display based on traffic streamlines and different functions, and then the park will be designed to be a public, interactive and popular science exhibition area. The film cultural park takes the film and television culture as the theme, taking the site of the Beijing Film factory as the dynamic point and radiating to the surrounding area. By combing the layout of the present building, the narrative axis of the landscape is obtained, and then the design of the site is perfect. The transformation design of the living area includes the Market green parking complex, commercial pedestrian street, ecological experience garden, memorial wood, military memorial places, community recreation Plaza, and a series of landscape nodes.

Finally, through the different scale of the prospective planning and design, the urban space is endowed with various functions of the contemporary urban life, and the urban green ecological network is constructed from the microcirculation of the city. The vitality of the region is effectively stimulated. At the same time, a healthy lifestyle, which is willing to stay in the open air and use green traffic, is gradually integrated into the lives of residents. The life of the people living in this area will become more beautiful, the characteristics of the north central axis will be better embodied, and the environment of downtown Beijing will be greatly improved.

生态绿网，活脉融城
北京市北中轴生态网络更新构建研究

Green Lines as City Veins
The Establishment of Ecological Networks around the North Central Axis of Beijing

曾筱雁、冯君明、李承玺、刘亚男、孟城玉、王宏达、王科
Zeng Xiaoyan / Feng Junming / Li Chengxi / Liu Yanan / Meng Chengyu / Wang Hongda / Wang Ke

研究区域是首都功能核心区向城市功能扩展区的过渡地带，也是城市北中轴线沿线范围，承担着延续北京轴线景观风貌和城市功能转变过渡的重任。

场地东起安定路，西至西土城路，北至北土城西路，南至黄寺大街、新康路，位于朝阳区、西城区以及海淀区三区交界之地。场地内既有元大都城垣遗址公园、蓟门烟树等历史文化节点，也有北京电影制片厂、中国科技馆等具有现代特征的元素。但是由于交通割裂、组团封闭等原因，现有的景观可达性较差。

此次规划设计以大连接与小循环，构建城市功能过渡地带生态绿网为规划目标，并结合生态绿网、活脉融城的设计理念，构建延续古城文脉的传统中轴景观。同时通过生态绿色网络、文化产业园区、便捷生活圈以及都市慢行系统的营造充分激发区域活力。

场地东起安定路，西至西土城路，北至北土城西路，南至黄寺大街、新康路，位于朝阳区、西城区以及海淀区三区交界之地，纵跨北三环，横跨北中轴线。北接中华民族园、奥林匹克体育中心，西邻元大都城垣遗址公园，南望地坛公园，场地面积约624hm^2。调研区域是首都功能核心区向城市功能扩展区的过渡地带，承担着延续北京轴线景观风貌和城市功能转变过渡的重任。

规划地块作为城市中心向自然的衔接过渡地带，对生态廊道的串联与延续有积极的意义，对于搭建区域绿色网络进而提升北中轴生态长廊的品质至关重要。

以立体绿化为触媒点，提高片区雨水的渗透力，缓解由城市带来的排水压力，将生态修复融入城市更新；在网络绿道的构建中，强调慢行优先，提高慢行舒适度，以各功能绿道为核心载体，与城市功能、轨道交通协同发展的慢行系统，推进绿色出行；并且以城市楔形绿廊为主干，以绿网串联各绿地为枝干，以生态化街道与线性绿地为毛细，将自然引入城市的各个角落。

在从平淡印象到城市记忆的打造中，首先以文化遗址为导向，加强历史地区及标志性建筑的保护，恢复场地、场所集体记忆，补充完善城市空间肌理与历史信息，延续区域风貌特色；其次在特色植被的规划中，依据地域特征，从生态角度出发，建设集特色植被展示、乡土植被保护为一体的区域植被景观。此外结合场地空间规划与多元文化，共同展现创意城市庆典活动，塑造独具魅力的城市名片，创造集体记忆。

在向人文活力乐园的转型发展中，首先以人本生活为主体，满足市民开展多样活动的需求，焕活区域消极空间与缝隙空间，用简单高效的手段实现对地面空间的合理划分，创造丰富的公共活动空间。在人文体验的塑造中，致力于场地文化的激活，寻求现代发展视野下包括绿色社区文化在内的多元文化融合。从活动组织、空间布局、设施安排三方面为市民提供精神公共乐园。此外建立现代智慧型数字体系，通过实现资源共享以及信息循环，使市民与场所相连，实现线上线下互动。

具体改造以宏观、微观结合为主，着力对区域内各种用地性质以及相应人群进行分析，通过对场地地域性分析可以得出：场地以居住用地为主，用地分布呈现单一趋势；商业与服务业用地分布不均匀，配套不全；在绿地与广场用地层面以现有公园绿地为主，社区游园、街边绿地等不足。

此外，区域人群以中老年为主，学生较多，白

规划愿景
Plan Visions

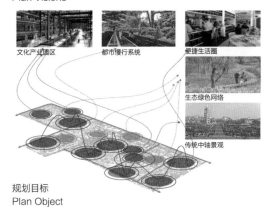

规划目标
Plan Object

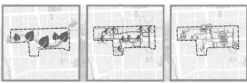

策略1 绿网整合

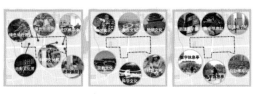

策略2 活力提升

策略3 文化渗透

设计策略
Design Strategy

规划设计 PLANNING & DESIGN

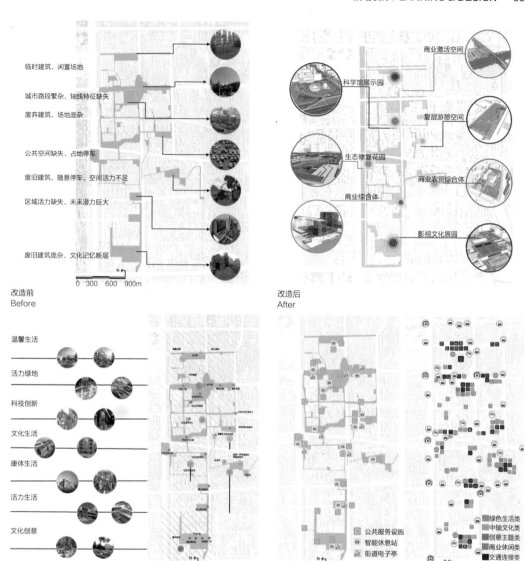

改造前
Before

改造后
After

功能定位
Function Design

信息设施
Information Facilities

交通设施
Traffic Facilicies

功能修补
Functional Improvement

领工作人士较少。因此，对于城市功能的活动空间需求较多。

在具体的规划设计中以完善区域功能为目的，多方面完善城市功能；在绿色交通设施方面，以绿色生活区块为主，为周边居民提供便捷的市民服务以及体育娱乐和室外休闲等服务设施，使居民生活更加便捷。

同时，中轴体验区结合多种类型的文娱设施，为当地居民和外来游客提供观赏景观、文化交流、体育娱乐等服务设施。科技创新区则是规划特色，该分区内设置多样化的科教文化、体育娱乐、休闲观展等服务设施。文化创意区提供科教文化、体育娱乐等设施，还设置开放的艺术展览区、景观区，增强场地的文化氛围。

结合场地内部休闲娱乐场地、运动场地、小卖部、公厕、公交站点等相关服务设施布局两级智能服务站点，满足居民及其他人群的使用需求。部分场地内部的智能服务站点除了基本的查询功能外，同时展示生态教育、科普及场地历史记忆等内容。

绿道规划建设的目的是通过绿道连接城市绿地，形成绿色网络，从而保护生态环境、提升城市形象，与城市的历史文化交相辉映。同时，绿道规划建设满足公平性原则，对本场地而言，由于场地交通系统层次复杂，绿道的可达性作为直接影响居民使用的因素占有重要地位。

综合绿道规划建设的生态性及公平可达性要求，基于空间句法进行都市型绿道选线。

基于现有场地交通系统规划，通过空间句法进行全局整合度分析，得出该场地内道路的可达性，其中可达性越高则线形亮度及暖度越高。将高亮的部分道路选出，作为绿道选线的基础。

在绿道选线的基础上，叠加北京市绿道整体规划、场地及周边绿地系统、景观风貌总体规划三个方面，得出初步绿道选线。

在初步绿道选线的基础上，通过对城市道路网络构成以及场地用地性质等因素分析，进行绿道选线修正，并与场地潜力地块进行叠加，得到最终绿道选线成果。

区域绿道的路径规划将公交站、地铁站等城市交通与区域慢行系统的连接进行修复与完善。同时，通过区块内现有以及新建的公共停车场与区域绿道网络相结合，根据区域功能分布相应增加自行车驿站的分布，完善区域内部机动车、自行车等相关车辆的服务系统。

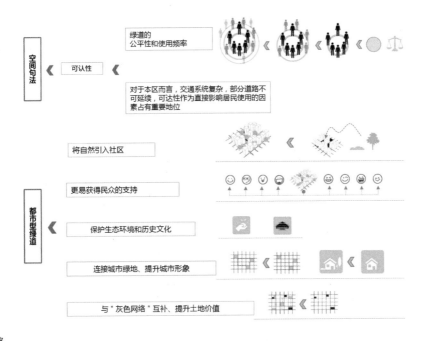

选线策略
Strategy

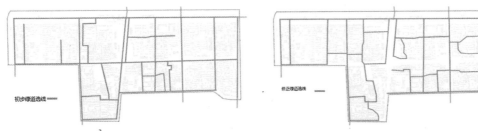

初步绿道选线
Preliminary Choice of Greenway

最终绿道选线
Final Choice of Greenway

区域绿道的改造主要用于串联区域功能，将各连接口（绿地连接口、居住用地连接口、商业用地连接口、教育科研用地连接口）与生活息息相关的城市功能进行串联，提升各功能连接点的路径可达性以及景观品质，构建慢行系统。

通过对场地内外城市环境的性质与特点进行分析，对地块内部的绿道网络进行整合与梳理，分为城市慢行路径、社区休闲路径、文化体验路径等，打造生态、健康、优质的都市慢行系统，鼓励整个区块内人群的绿色出行。

对场地内可以改造提升的区域进行适度开发，构建生态绿色建筑、智能绿色交通、生态网络系统，对重点区域进行生态保护，推进生态科普，打造海绵城市、智慧城市。

依据场地用地性质以及场地现存植被状况，将其划分为 4 个植被主题区域：特色种植区、绿色园艺区、乡土植被区和商业景观区。

运用海绵城市建设相关手段，打造绿色屋顶、生态滞留设施、渗透沟、雨水花园等雨洪管理设施，构建生态城市。

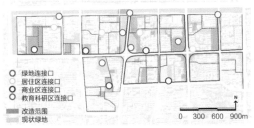

绿道接入点
Greenway Access Point

停车场及自行车驿站
Park and Bicycle Station

生态分区
Ecology Partition

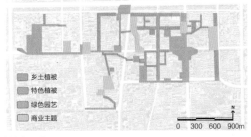

植被主题分区
Vegetation Theme Partition

雨水收集策略一
Rainwater Collection Strategy

雨水收集策略二
Rainwater Collection Strategy

102 规划设计 PLANNING & DESIGN

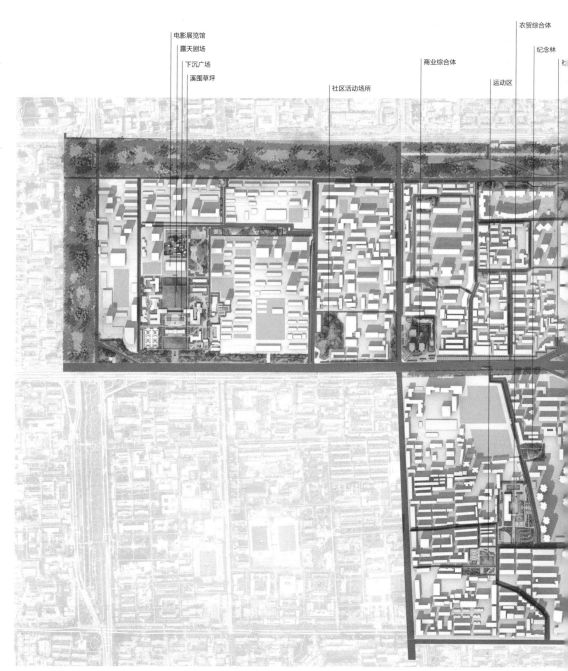

规划总平面
Master Plan

规划设计 PLANNING & DESIGN

1. 北中轴街道改造

场地位于北京中轴线北段,从鼓楼外大街至北辰路,直抵元大都城垣遗址公园。街道两边主要分布 20 世纪 70～80 年代居住区、企业单位以及少量绿地和商铺。场地总面积约 15.2hm^2。

经实地调研和采访,参考国内外城市中轴设计分析发现,该路段轴线定位模糊,一般城市快速路景观为主,中轴特色不鲜明。同时,沿街绿地利用不充分,缺乏公共使用功能,环境亟待恢复与提升。

为建设具有标志性的中轴景观,彰显城市形象,方案将北辰路一段主车道下穿,盖板建绿,延续从奥林匹克森林公园至元大都城垣遗址公园方向的绿廊,从自然引绿至主城区,延伸城市绿肺。

方案将场地从南至北分为城市生活段、街角公园段、中轴绿地段,以叙事性手法,分为回顾、发展、波折与复兴四个部分。其中鉴史广场是以铜板浮雕墙、文化柱等元素塑造的纪念性城市广场。水阶荡池以下沉地形、流动水台阶、富有节奏变化的规则树篱等表达历史发展之动势。多处街角空间以微地形、树阵和绿道等形式改造为城市缝隙绿地。

城市生活段街景模式
Streetscape Model of City Life Section

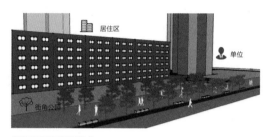

街角公园段街景模式
Streetscape Model of Corner Park Section

中轴线街道景观改造平面
Design Plan of Axis Street

规划设计 PLANNING & DESIGN   105

鉴史广场效果
Perspective of History Square

街角公园效果
Perspective of Corner Park

水阶荡池效果
Perspective of Waterscape

鸟瞰
Aerial View

1 乡野草甸	6 镜面水池	11 街心花园
2 覆土建筑	7 下沉通道	12 健康步道
3 模纹花坛	8 科技中心	13 绿道
4 水台阶	9 中轴塔	14 谧园
5 鉴史广场	10 康体花园	

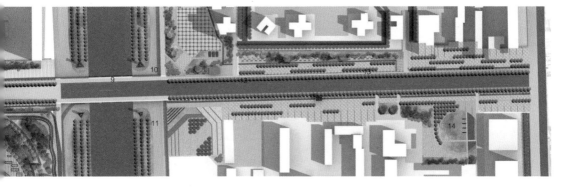

2. 中国科技馆周边更新

场地位于北京市北中轴西侧，西邻京藏高速，东邻北辰路，北接北土城西路，南侧紧邻北三环。场地东侧、西侧均是居住用地，南侧和北侧分别紧邻商业、办公和展览用地。

场地面临的主要问题有：公共开放空间被占用，难以发挥更高效益；公共服务不健全，服务于市民的公用设施有待完善；慢行系统混乱，慢行体验差；空间封闭，缺乏人性化体验。

因此，采取连接、共享、存量更新的策略。连接即空间的连接，活动、交流的连接，社会的连接；共享即服务共享、空间共享、信息共享；存量更新是指寻找逝去的角落，更新破旧的空间。

在此策略的基础上，进一步确定了公共空间共享、线性空间连接、绿道网络构建的目标，依托城市道路与公共开放空间共同搭建起集步行、骑行、休闲、活动于一体的绿道网络。

选择四处具有潜力的地块进行设计，分别是科学中心绿地景观、科技绿廊带状活动区、社区活动中心和企业园区开放绿地。科学中心绿地是以科学展示和科学普及为主题的绿地；科技绿廊是一系列科技创新、文化服务等办公企业的附属休闲带，以突出现代感、科技感为特色，以连接、贯通、激活、融合为目标；社区活动中心是位于社区内部的活动空间，主要为老人及儿童服务；企业园区开放绿地是位于一些企事业单位附近的小型活动空间，提供室外休闲和交流座谈的场地。

慢行系统策略
Strategy of Slow-traffic System

公共空间策略
Strategy of Public Space

绿道策略
Strategy of Greenway

科学中心区域鸟瞰图
Aerial View

规划设计 PLANNING & DESIGN

① 主入口
② 入口草坪雕塑
③ 核心广场
④ 广场跌水小品
⑤ 广场旱喷景观
⑥ 展览馆台地景观
⑦ 科普林
⑧ 地铁站出入口
⑨ 展览馆A
⑩ 展览馆B
⑪ 活动广场
⑫ 跌水台阶
⑬ 次入口
⑭ 自行车停车驿站
⑮ 林下活动场地
⑯ 空中栈道
⑰ 科学原理体验区
⑱ 雨水花园
⑲ 条石驳岸景观
⑳ 双轮车障碍体验
㉑ 攀登架
㉒ 拉起车体验
㉓ 激光轨迹体验
㉔ 宇宙之奇
㉕ 生命之秘
㉖ 声音之韵
㉗ 地质之奥
㉘ 四大发明展示区
㉙ 咖啡活动场地
㉚ 办公区休闲场地
㉛ 内部停车场

科学中心平面
Design Plan of Science Center

落水景观效果
Perspective of Waterscape

咖啡小筑效果
Perspective of Cafe

① 科技馆区域开放绿地
② 慢行步道体系
③ 办公区休闲空间
④ 办公区活动广场
⑤ 儿童活动场地
⑥ 居民健身广场
⑦ 居住区休闲廊架
⑧ 商业入口广场
⑨ 树阵广场
⑩ 办公区开放空间
⑪ 玫瑰园
⑫ 街旁小广场
⑬ 跌水广场
⑭ 林下休息
⑮ 镜面水池
⑯ 下沉广场

科技馆周边设计平面
Design Plan of Science Museum Area

3. 影视文化创意园改造

在上位规划中，该片区定位为影视产业艺术区，依托现有元大都城垣遗址公园绿地，发掘场地潜力，通过活力点的打造，构建区域绿网，并与东侧绿网衔接。

设计中以影视企业文化与场地特征为依托，将北京电影制片厂（以下简称"北影厂"）作为区域绿网内核，进而向四周辐射带动区域活力，建立一个具有循环经济手段、公众参与功能、绿色生态展示、舒适轻松氛围的影视文化创意园。

通过对场地现状的调研，针对区域特征弱化的挑战，我们将影视文化作为区域的设计主题，以北影厂址为活力点向四周辐射确定最终的设计范围。通过梳理现状建筑布局情况，得出景观叙事轴线进而完善场地设计。

该片区方案生成分为三步，分别是保留、唤醒和激活。北影厂作为历史文化场所，首先要确定建筑的价值并提出针对性的改造措施，其次通过对中国电影发展大事年表的梳理，并借鉴电影常用的叙事手法，在园区设计景观叙事轴线，让人们对场地的历史有更深切的体会。轴线共分为四部分：分别代表着开端、发展、高潮和尾声，同时在空间处理上采取开合对比的方式，增强体验感。最后通过新功能的引入为园区注入新的活力，主要是对原有建筑的功能进行整合划分，增加了配套服务、企业办公和对外展示三个功能。

地块范围 Range　主题选择 Theme　活力点确定 Vitality Point

周边交通分析 Traffic　场地辐射方向 Radiant Range　客观阻碍分析 Hindrance Factor

最终设计范围 Design Range　保留建筑分析 Reserved Building　景观轴线生成 Axis of Landscape

设计策略
Design Strategy

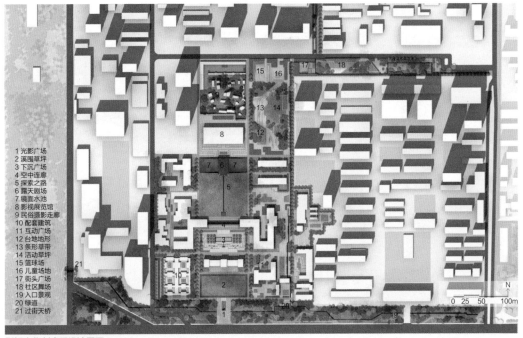

1 光影广场
2 溪围草坪
3 下沉广场
4 空中连廊
5 探索之路
6 露天剧场
7 镜面水池
8 影视展览馆
9 民俗摄影走廊
10 配套建筑
11 互动广场
12 台地地形
13 条形草带
14 活动草坪
15 篮球场
16 儿童地地
17 街头广场
18 社区舞场
19 入口景观
20 绿道
21 过街天桥

影视文化创意园设计平面
Design Plan of Film & Television Cultural Park

规划设计 PLANNING & DESIGN 109

绿道系统分析
Greenway System Analysis

下沉广场效果
Perspective of Square

沿街带状绿地分析
Greenbelt Analysis

摄影走廊效果
Perspective of Cultural Street

园区绿地网络分析
Green Network Analysis

企业广场效果
Perspective of Square

影视文化创意园鸟瞰
Aerial view

4. 商业综合激活

商业综合体区域基于场地特点以及对未来潜力分析，以城建大厦为主体进行商业综合体改造，通过对主体建筑的功能转换、商业附建的建设以及外环境改善等的结合打造公共商业环境，此外在功能布局中，以满足多种商业功能需求为出发点整体布置。同时，景观构筑物、铺装、屋顶花园等外部环境要素均整合布置，塑造多样、协调、有体验序列的商业活动空间。

1 地下车库入口
2 屋顶花园（附建）
3 商业附建
4 旋转喷泉
5 空中走廊
6 屋顶花园（主建）
7 商业主入口
8 树池座椅
9 地铁站入口

功能分析
Function Analysis

交通分析
Traffic Analysis

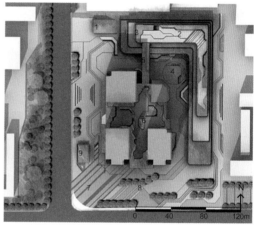

商业综合体设计平面
Design Plan of Commercial Center

商业综合体效果
Perspective of Commercial Center

规划设计 PLANNING & DESIGN 111

生态体验园区位于健德桥附近，园内水循环被现状硬质盖板阻隔，建筑密度大且缺乏开放空间。以水生态的修复为出发点，恢复地下水流，并结合台地高差进行汇水调整，提升外环境，塑造生态休闲体验空间。并且从生态展示、观赏游憩两个层面布置户外空间。

在功能布局中，以为居民提供良好的公共空间为目的，完善城市开放空间。与此同时，以休闲、游憩功能为依托，恢复该节点的生态价值。

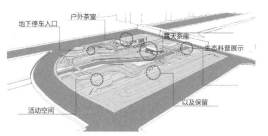

功能分析
Function Analysis

交通分析
Traffic Analysis

1 滨水花台
2 元大都水系
3 滨水平台
4 地下停车场入口
5 台地花台
6 城垣遗址
7 户外茶座
8 地铁站入口

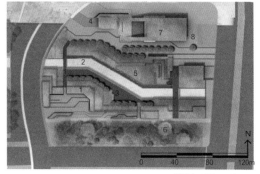

健德桥周边设计平面
Design Plan of Jiande Bridge Area

河道修复效果
Perspective of River Restoration Area

5. 活力生活圈设计

该地块在上位规划中属于活力生活圈，其规划定位是提升居民生活品质、打造优质居住环境、实现联系你我的绿色活力生活圈，主要目标是为周边居民建立富有活力的休闲区域，并依据其优越的地理位置建立北京市中心现代的生活圈。此区域周边交通、绿地资源丰富，通过绿廊可以连接居民生活区，向北可连接马甸公园、元大都城垣遗址公园等绿色资源。

基于现状分析，总结出该区域面临的困境和机遇，并提出四个对应的解决策略：废弃破旧场所—拆改、具有景观基础的场地—提升、景观较好的场地—保留、绿廊慢行串联。整个方案由活力邻里区、纪念林活动区、综合广场活动区和绿廊四个部分组成。

活力邻里区以简洁景观为主，主要服务周边居民，在更新时注入了儿童游戏、草坪剧场、健身广场等功能。

纪念林活动区以松柏景观为主，沿原有道路设置了景墙，并将原有道路细致化；依据原有场地进行人性化设计和景观化处理，形成林荫广场、林中漫步、荫下小憩等不同的功能区，营造出一个具有艺术气息的场所。

综合广场活动区以新型综合体为主，围绕其营造休息景观，为周边居民提供更多的活动空间，并与商业空间相互连接，提升该区的生活质量。

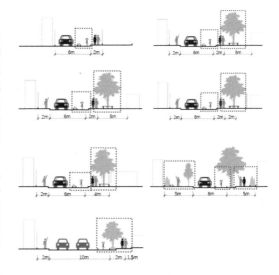

绿廊剖面分析
Analysis of Greenway

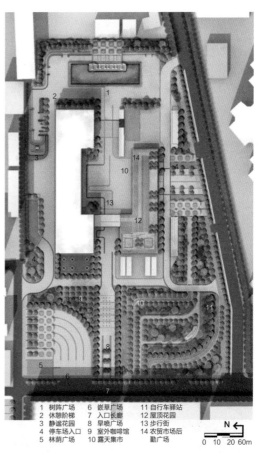

新型农贸综合广场设计平面
Design Plan of Market Plaza

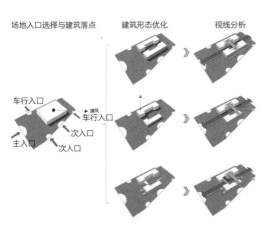

新型农贸综合广场方案生成分析
Analysis of Market Square Plan Generation

规划设计 PLANNING & DESIGN

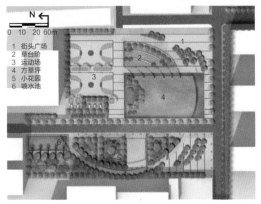

邻里空间设计平面图
Design Plan of Neighborhood space

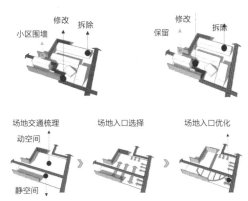

活力邻里方案生成分析
Analysis of Vigorous neighborhood program generation

纪念林设计平面图
Design Plan of Memorial Wood

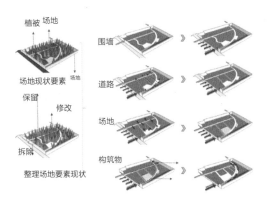

纪念林方案生成分析
Analysis of Memorial Forest Program Generation

农贸广场效果
Perspective of Agricultural Trade Plaza

邻里生活区效果
Perspective of Neighborhood Living Area

纪念林效果
Perspective of Memorial Wood

活力生活圈鸟瞰
Aerial View of Vital Life Circle

6. 文化生活圈

文化生活圈地块北临中国科学技术馆、玫瑰园、马甸公园，隔一个街区有元大都城垣遗址公园，南有西黄寺。整个场地南北侧连通需求较高，文化底蕴深厚。设计方案以此为切入点，形成以绿道纵向连通、横向成网的整体策略，并串联以文化功能和生活功能为主的绿色节点与基础设施，形成符合片区特征的绿色生态生活网络。

绿色停车综合体现状为临时停车场，是附近最大的一块公共空间。设计方案使用羊毛线模型对人群路径进行自组织模拟，得到自发产生的综合路径作为场地骨架。通过抬升台层，来解决停车需求与公共游憩活动需求的矛盾、机动车穿行与步行穿行的矛盾，并且在台层边缘设置台阶草坪和商业生活建筑，从而激活片区活力，满足居民生活需求。

军事外交纪念场所现状为废弃建筑与工厂建筑，旨在重新恢复附近已关闭的军事外交纪念场所。设计方案主要通过改造工厂建筑作为陈列馆，并使用水景的变化表达我国外交的历史过程。

西黄寺作为藏传佛教学院所在地，现状周围有围墙封闭，隔绝了文化的交流。设计方案将西黄寺分为对内教学区和对外交流区。对内教学区保持封闭的教学环境，对外交流区将围墙打开，连通寺内外文化。

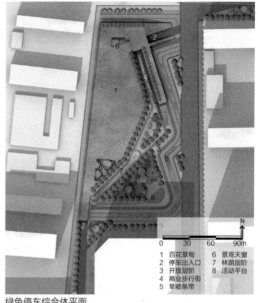

绿色停车综合体平面
Plan of Parking Complex

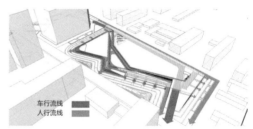

绿色停车综合体立体交通解析
Traffic in Parking Complex

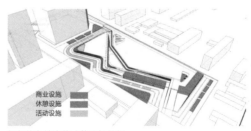

绿色停车综合体功能设施解析
Function in Parking Complex

军事外交纪念场所平面
Plan of Memorial Plaza of Military Diplomacy

设计策略
Design Strategy

规划设计 PLANNING & DESIGN   115

停车综合体鸟瞰
Aerial View of Parking Complex

纪念场所效果
Perspective of Memorial Plaza of Military Diplomacy

停车综合体效果
Perspective of Parking Complex

绿道效果
Perspective of Greenway

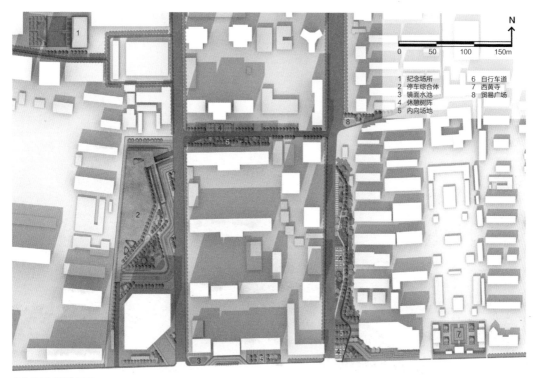

1 纪念场所
2 停车综合体
3 镜面水池
4 休憩树阵
5 内向场地
6 自行车道
7 西黄寺
8 贸易广场

文化生活圈地块平面
Plan of Culture & Living Area

7. 安贞路－安华路社区生活圈

安贞路－安华路社区生活圈以安贞路－安华路街心绿地、安贞西里社区废弃用地为依托，场地北接元大都城垣遗址公园，南接柳荫公园，周边区域以居住社区为主。

安贞路－安华路街心绿地区域基于场地特点以及对周边用地性质分析，以街心绿地为主体进行功能多样化改造。通过对街心绿地的功能转换打造社区休闲生活环境。在功能布局中，以满足场地周边社区居民的需求为主要出发点整体布置，塑造多样、协调、有体验序列的社区休闲生活空间。

安贞西里社区废弃用地不能满足社区居民的需求，且与社区氛围不相符合。以废弃用地景观提升改造为出发点，将打造集休闲活动、文化娱乐、社区生活为一体的社区休闲广场，提升景观环境，塑造生态休闲体验空间。

遵循以人为本、绿色生态的设计理念，以服务周边社区居民及工作人员为目的，解决场地现有问题，对场地功能进行整合，连接周边绿地形成绿色生态网络，打造温馨的社区生活氛围。

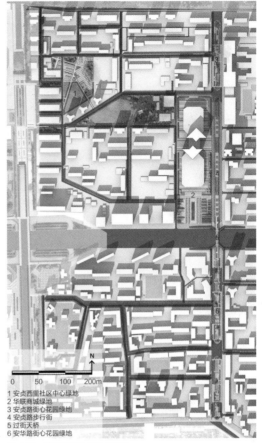

1 安贞西里社区中心绿地
2 华联商城绿地
3 安贞路街心花园绿地
4 安贞路步行街
5 过街天桥
6 安华路街心花园绿地

安贞路－安华路区域设计平面
Design Plan of Anzhen Street and Anhua Street Area

过街天桥效果
Perspective of Overpass

规划设计 PLANNING & DESIGN   117

STEP 1 确定场地主体建筑社区活动中心位置，分析场地周边交通流线及与人流方向，得出场地与周边的接入点。

STEP 2 为方便步行交通，连接各接入点选取最短路径。

STEP 3 对得出的最短路径进行整合，得出最终场地道路，并根据场地功能需求布置休闲林荫广场，沉水花园，露天剧场、活动大草坪。

形态演绎
Morphological Deduction

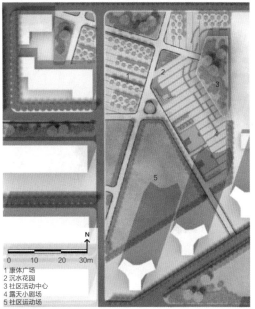

1 康体广场
2 沉水花园
3 社区活动中心
4 露天小剧场
5 社区运动场

社区中心绿地平面
Design Plan of Community Center Green Space

模式一
Mode Chart One

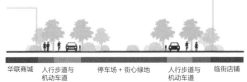

模式二
Mode Chart Two

下沉剧场效果
Perspective of Sink-style Theater

步行街效果
Perspective of Pedestrian Street

园艺花园效果
Perspective of Horticulture Garden

康体乐园效果
Perspective of Recreation Garden

The north central axis of Beijing plays an important role in urban planning. It is important to connect the old district with the new district in the city, to connect the history and the nature, to affect the quality of the city and the well-being of the residents.

The design site is across the North three ring road. It is south to Huangsi Street, the New Kang Road, north to the North Tu Cheng Road, west from the West City Road, and east to the Anzhen road. It is among the urban functional development zone. This site is the junction area of Dongcheng, Xicheng and Haidian district. It was designed for dwelling mainly, with office function as a supplement, scattering with many unit yard community and closed community built in the 1980's of last century. Although the plot contains the central axis, in recent years, the development and update speed was relatively slow. The updating of the area is an important part of shaping the characteristics of north central axis. It will help to clear the function and position of north central axis, to inherit the history and culture of Beijing, to improve the ecological environment of the area, and to improve the quality of people's life.

In order to achieve the goal of constructing "micro-circulation", we choose the most suitable way of regional status-sharing the neighborhood life, activating the block with its core. Firstly, neighborhood Life Circle is chosen as Neighborhood Green Point, and green network skeleton is connected, then the city Green net will be updated. Next, the choice of the green point of the neighborhood committee also has certain basis, namely the existing green space, the demolition and reconstruction land. The vacant open space is identified as the neighborhood committee green space. Finally, the Green Point of neighborhood will be built to be a neighborhood life circle, to develop the green network skeleton, so as to get through the microcirculation system, and then to form a regional green inheritance.

Referring to the system planning of Beijing Central city Greenbelt, the urban fringe greenbelt will be penetrated into the inner city like a wedge. Beijing Yuan Dadu Ruins Park undertakes the important role of connecting link. Therefore, Beijing Yuan Dadu Ruins Park will be the green center of the region of planning area. It will expand to the south, the slow traffic system will be built, and the construction will be focused on the green center of this region. Combined with the function orientation of new town and old town, Haidian and Chaoyang District, a series of renovation tactics will be put forward to the Beijing Yuan Dadu Ruins Park. First is to eliminate the enclosing wall to create space. Then is the demolition of pro-construction and remodeling function. Finally, the area of the transformation and upgrading of filter will be determined according to the above positioning and strategy, contributing to the Green Core of great communication and the Green Point Microcirculation neighborhood.

The selection of green core is mainly based on the present situation and future potential. In 2020, three new metro lines will be constructed: Line 9, Line 12 and Line 19. With four new subway stations, there will form a new Urban Subway Transit Network—"Two-horizontal & Three-longitudinal".

To build the green ecological network, the public transport system will be combined with the urban green network construction as much as possible. By this way, the accessibility of green space will be enhanced and green network could be shared by all people. Based on convenient connectivity, the framework of the regional green network is constructed and combined with the current traffic system. The current road with potentials will be chosen to be developed into the green skeleton. Finally, according to the existing properties and characteristics of the site, the theme of each green net skeleton is planned.

Based on the analysis of current situation, this plan will focus on the renewal of urban green network, taking the construction of green heart, regional green center and the neighborhood green center as the planning strategy. Good communication and micro-circulation will be considered as the effective Connection. As a result, green, living and sharing network system will be established. Finally, through green renewal, green space in the city could be used for public communication and outdoor activities, so that people can feel nature at anytime and anywhere. In the future, the renewal of green land will drive the peripheral region to continue to develop, and will inject more vitality into the area.

向"芯"生长
北京北中轴绿网更新
Centripetal Growth
Renewal of Beijing Central Green Network

于雪晶、王美琳、皇甫苏婧、赵人镜、王宇泓、石渠、张真瑞
Yu Xuejing / Wang Meilin / Huangfu Sujing / Zhao renjing / Wang Yuhong / Shi Qu / Zhang Zhenrui

　　北中轴在北京城市规划中有重要的作用，是承接古城与新城、连接历史与自然的重要脉络，影响着城市品质及人民福祉。

　　场地横跨北三环，位于东城区、西城区、海淀区的交汇处。南至黄寺街，新康路，北至北土城西路，西至西土城路，东至安贞路，位于城市功能开发区之内。场地内以居住区为主、办公功能为辅，20世纪80年代建成的许多单位社区和封闭社区散落其中。尽管区域中包含中轴线，但区域的发展和更新速度缓慢，已不能满足北中轴日益提升的城市印象塑造诉求。

　　在分析区域现状的基础上，此次规划将以城市绿网更新为核心，以建设生态绿心、区域绿色核心区、邻里绿色中心为方法，以良好的沟通和微循环为手段，构建绿色生活共享网络体系。

在北京中心城绿地系统规划中，城市外围绿地以楔形绿地形式向内城渗透，元大都城垣遗址公园承担着承上启下的作用。因此，将元大都城垣遗址公园作为规划范围内的区域绿芯，向南扩展，构建慢行体系，结合新城、旧城、海淀、朝阳各区功能定位，对元大都城垣遗址公园提出以下改造策略：①消隐围墙，将现有的闭塞空间转化为积极活泼的空间；②拆除临建，重塑功能。将现状功能混乱的临建拆除，赋予其文化创意、运动康体、活力商业等功能。最终根据以上定位及策略，提出"向芯生长"主题，确定区域绿芯的改造及提升范围，打造生态绿心、功能核心和区域中心，为联通街区绿核、微循环邻里绿点做基础。

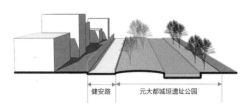

现状问题一：封闭堵塞的空间

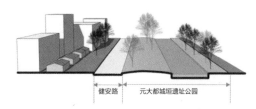

现状问题二：功能混乱的临建

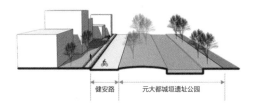

解决策略一：消隐围墙，营造空间

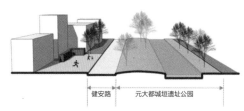

解决策略二：拆除临建，功能重塑

改造策略
Transformation Strategies

围墙空间分析
Reform Wall Analysis

区域重塑分析
Reform Section Analysis

核心空间及潜力地块分析
Development Potential Areas Analysis

街区绿核规划
Block Greencore Planning

区域绿网骨架构建规划
Regional Green-network Construction Planning

街区绿核的选取主要是依据现状及未来潜力。到 2020 年，场地内会扩建 9 号线北延、12 号线、19 号线三条地铁线路，形成"两横三纵"的城市轨道交通网络。设计所构建的绿色生态网络，将与公共交通体系尽可能复合，增强绿地可达性，实现绿网共享。在对应场地内新增四处地铁口，两处与绿地相接，另外两处将纳入区域的绿核体系，重点打造。基于便捷连通性，构建与现状交通体系复合的区域绿网骨架。在现状道路基础上，选择出潜力路线，综合路宽、绿化、建筑距离等因素，排除现状较差、改造难度较大线路，选择面向更多小区出入口的道路，结合街区绿核构建复合的绿色慢行体系。最后，根据场地现有属性及特征，对每条绿网骨架进行主题策划。

为达到构建"微循环"的目标，选取最适宜区域现状的更新方式——共享邻里生活，以其为芯，激活街区。

邻里中心的实质是将多种生活服务设施集中布置的"商住绿色综合体"，是家庭生活在绿色环境中的外延，是所在邻里街区的核心。其服务人数在 6000～8000 人之间，服务半径 800～1200m，与现状被城市主次干道割裂形成的街区吻合。

邻里中心摒弃了传统住区零散的商业和服务模式，将服务集中在核心生活圈内。以绿地为载体，赋予其邻里服务功能，激活邻里，改变封闭的生活方式。选择邻里生活圈作为邻里绿点，连接绿色网络骨架，更新城市绿网。

第一步，对区域内的封闭小区进行改造，采用的主要策略有：①边界彻底打开；②居住区拆分为居住小区；③红线公园景观改造。其次，将具有改造潜力的现状绿地、拆除重建用地、闲置空地确定为邻里绿点。

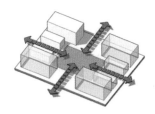

策略一：边界彻底打开

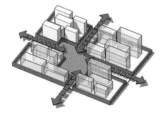

策略二：居住区拆分为居住小区

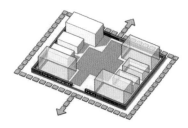

策略三：红线公园景观改造

邻里绿点规划策略
Green Point Planning Strategy

第二步，构建向绿网骨架生长的邻里生活圈。主要采用因子叠加的方法，选取的因子有：①功能混合度，即在一定区域内不同土地类型的相对临近性，反映在居住、商业、办公等不同功能用地在空间中的合理混合分布的程度；②人口密集度，该因子一定程度上反映人口基数；③边界可利用程度。围栏边界可相对软化的边界易于改造，且利用潜力较大。

最终，结合邻里绿点构建邻里生活圈，向绿网骨架生长，自此打通微循环体系，构成区域绿色血脉。

功能混合度
Functional Diversity Analysis

人口密集度
Population Density Analysis

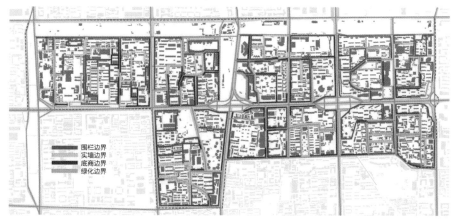

大院边界可利用程度
Boundary Utilization Analysis

邻里绿点选择
Neighbourhood Green Point Analysis

规划设计 PLANNING & DESIGN   123

邻里生活圈规划范围
Neighbourhood Circulation Planning

影视复兴廊道意向
Movie Revival Corridor Perspective

马甸康体廊道意向
Sporting Corridor Perspective

科技创意廊道意向
Technological Creative Corridor Perspective

古今传承廊道意向
Historical Corridor Perspective

乐享生活廊道意向
Entertainment Corridor Perspective

邻里绿点意向
Neighbourhood Green Point Perspective

124 规划设计 PLANNING & DESIGN

规划总平面
Master Plan

规划设计 PLANNING & DESIGN   125

草坡剧场 | 停车场 | 元大都城垣遗址公园 | 街旁绿带 | 地下车行入口 | 华联商厦前广场 | 街心公园集中区

科技创意广场 | 居民生活广场 | 展览馆及前广场 | 古今传承中轴绿地 | 乐享生活绿廊

规划设计 PLANNING & DESIGN

1. 影视复兴廊道片区改造——北京电影制片厂规划设计

北京电影制片厂（以下简称"北影厂"）位于影视复兴廊道区域内。目前存在的问题主要有两个方面：①北影厂与周边用地的连接程度较差；②北影厂搬迁后，遗留的厂址保护与利用较差，景观品质有待提升。规划目标是复兴影视历史与文化；建立综合电影产业、教学、研究的示范基地；打造具有吸引力的、以电影为主题的景观，激活场地活力，重塑北京电影厂历史。

地形方面，塑造个性地形，将电影历史不同阶段的表现结合场地个性地形处理，形成高低错落的文化纵轴；艺术景观结合下沉地形，形成多功能的景观横轴等。建筑方面，拆、保并行，重置建筑功能。还原影视文脉，保存具有历史记忆的建筑；延续中轴骨架，新建满足现代需求的建筑。道路空间方面，内、外联系，突显绿核品质。

总体而言，规划路网，为大连通做底；激活中轴，为微循环打基；强化空间，为微循环筑础。公共开放空间统筹兼顾，附属绿地内外结合；纵、横绿轴个性突显；围合绿地品质提升。使之成为功能与景观相结合、地形与文化相结合、风格与技术相结合的街区绿核。

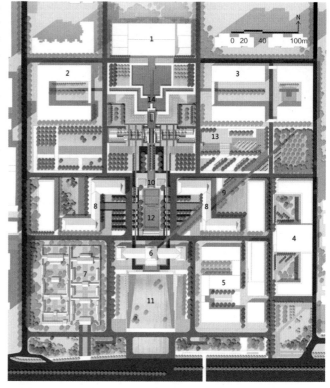

1 影视放映馆	4 影视创作	7 古装戏体验王府	10 历史文化轴	13 下沉横轴
2 影视教学	5 影视周边买卖街	8 多功能办公楼	11 主楼绿毯	14 中心舞台
3 影视科研	6 影视历史展览馆	9 "汲古铸今"廊	12 中轴水景	

北京电影制片厂规划平面图
Design Plan of Beijing Movie Factory

策略1：规划结构

策略2：保存建筑

策略3：新建建筑

策略4：规划路网

策略5：激活中轴

策略6：强化空间

策略7：附属绿地内外结合

策略8：纵横绿轴个性突显

策略9：围合绿地品质提升

改造策略 Reform Strategies

规划设计 PLANNING & DESIGN

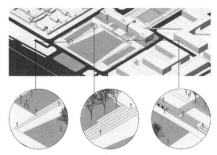

主楼绿毯区　Perspective of Green Space

中轴水景效果　Perspective of Waterscape

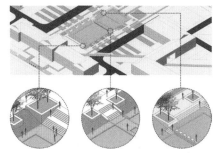

中轴水景区　Perspective of Waterscape

下沉横轴效果　Perspective of Axis

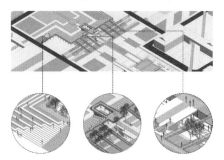

下沉横轴及高架舞台区　Perspective of Axis

高架舞台效果　Perspective of Elevated Stage

鸟瞰　Aerial view

2. 马甸康体廊道片区改造

此区域为马甸康体区，周边有元大都城垣遗址公园、马甸公园、双秀公园等现状公园，绿化环境较好。马甸公园使用现状良好，设计仅对北侧与北土城公园临近边界进行改造。双秀公园现状条件较好，整体保留，仅加入部分运动设施小品，呼应马甸康体区域定位。元大都城垣遗址公园作为规划层面中的区域绿芯，拥有极佳的绿化环境。但在京藏高速与北土城西路交界处道路交叉分割，连续性在此被打断，设计将重新规划此区域，以连接北土城及马甸公园，构建连续的绿色廊道。

设计将拆除区域内所有建筑，完全保留南侧现状绿地，设置立体廊道打通南北两侧关系。设计延伸入北土城公园及马甸公园，构建连续、统一的公园体系。由空中廊道划分出内外广场，外广场以通行、集散功能为主，内广场以休闲、户外茶室功能为主，动静有序。地铁前广场通过绿篱围合出一个半开敞空间，作为从地铁站出入口的前序景观，提供户外等候停留的场所。

除此之外，还设有景观式自行车停靠站，避免自行车随意摆放。地下商业及停车场连接地铁站，打造集餐饮、售卖、集散于一体的绿色建筑。空中廊道将从现状地形上方穿越，连通两侧绿地，避免人为接触地面，最大限度地保护城墙遗址。结合铺装、廊架、座椅、汀步等等，产生丰富的高差变化，并利用高差设置跌水景观。

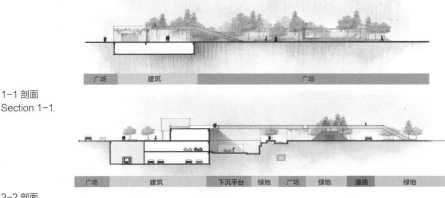

1-1 剖面
Section 1-1

2-2 剖面
Section 2-2

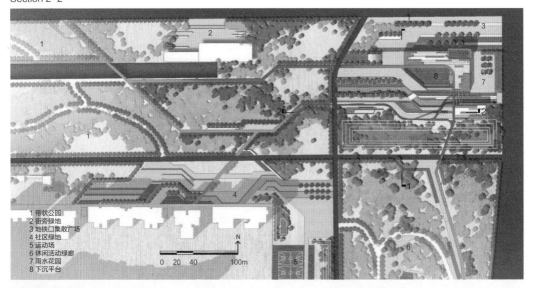

1 带状公园
2 街旁绿地
3 地铁口集散广场
4 社区绿地
5 运动场
6 休闲活动绿廊
7 雨水花园
8 下沉平台

马甸片区设计平面
Design Plan of Madian Area

规划设计 PLANNING & DESIGN   129

鸟瞰
Aerial view

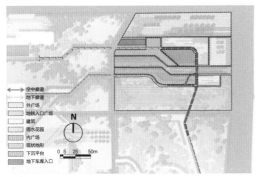

功能分区分析
Functional Division Analysis

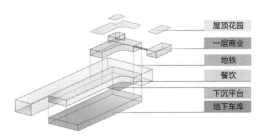

建筑结构分析
Building Structure Analysis

地铁出入口效果
Perspective of Subway Entrance

空中廊道效果
Perspective of Air corridor

跌水景观效果
Perspective of Waterfall Landscape

内广场效果
Perspective of Inside Square

3. 科技创意廊道片区改造

科技创意廊道片区位于规划场地中部，西起京藏高速，东至裕民中路，北临北土城西路，南至黄寺大街。在上位规划的"大连通·微循环"体系中，结合区块功能混合度、人口密度及周边小区边界类型等因素选定该区域为大连通特色廊道之一。同时结合上位规划，对该区域重点地块提出改造意向：打造创意街区景观、趣味市民花园。整体上形成连接南北的绿色廊道，带动周边区域活力。

规划提出两点目标：①以最大化共享绿色为导向，开放空间、提升品质；②以多样化场地空间为平台，激发活力、促进交流。

北部地块主要在商务办公的大环境中，充分考虑人群交流的需求。从以下方面进行改造：①增加可便捷到达的绿色共享空间，同时增加创意性元素，为上班族提供户外休闲及交流的可能。②将办公建筑周边围栏拆除，改为绿化边界的同时植入创意性元素。③整合并充分利用现状停车场，以解放建筑周边空间，提供更多活动的可能性。④对整合之后的空地及现状绿地进行再设计，增强可进入性，增加绿色交流共享空间。⑤增加部分建筑屋顶绿化，提升环境品质。拆除玫瑰园边界围栏，通过绿化进行适当隔离。⑥结合现状场地使用情况，对部分硬质铺装进行改造，以达到更好的使用效果。⑦南北连接的地下通道，延长入口空间，增加无障碍设计。⑧增加涂鸦文化，增强场地的生活气息与活力氛围。

对南部地块，综合考虑周围以居住用地为主的特征，着力打造社区邻里小花园性质的绿地空间。在场地中构建"盒子空间"，增强趣味性。首先，打破现有围墙边界，增强社区与场地的交通及景观联系。其次，沿城市道路一侧增加绿色开放空间，改善慢行交通体验。接着，在种植空间上注重环境的呼应，使北部一系列的绿色共享空间和南部小花园的盒子空间，整体串联成科技创意廊道。

1 带状公园
2 阳光草坪
3 休闲娱乐场地
4 户外活动场地
5 创意集市
6 儿童活动场
7 粗质花田

科技创意廊道片区平面
Design Plan of Technological Creative Area

景观结构
Landscape Structure

策略一：拆除围栏边界

策略五：绿化围栏边界

策略二：增加绿色共享空间

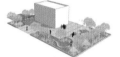

策略六：改善人行环境

策略三：整合临街停车位

策略七：增加地下通道无障碍设施

策略四：增加屋顶绿化

策略八：打开小区围墙边界

改造策略
Reform Strategy

儿童活动场效果
Perspective of Children's Playground

休闲娱乐场地效果
Perspective of Recreation Area

活动小剧场效果
Perspective of Outdoor Theater

户外活动场地效果
Perspective of Outdoor Activities Ground

4. 古今传承廊道片区改造——北京中轴线

北辰路位于中轴规划中的时代精神段，集中体现北京现代化城市风貌，现状与上下段沟通差，轴线感弱，南北向缺乏大块公园绿地衔接。

本次设计提出"下沉城市道路，建设中轴公园"的概念。旨在强调中轴线的仪式感，将其所承载的历史记忆重现，将其割裂的空间缝合，使城市道路消隐于自然和绿色之中，这也将是未来城市发展的方向。

为创造尽可能多的绿色空间，将道路下沉盖板，起翘延续，并对交通进行梳理。改造后将机动车道下沉，小区及岔路车行使用地面辅道。公交车站迁移，保证交通及人行顺畅。保留行道树和分车带种植，将人群活动向心聚集，提高区域活力。

在交通骨架基础上，将公园由内而外分为活动、慢行、休闲三类条带。充分考虑周边现状，在重要节点扩展延伸中央活动带，形成公园的整体结构。设计延伸到北土城公园内，采取轻巧的栈桥结构搭接到现有城墙硬质铺装上，将慢行体系与北土城相连，公园也成为渗入城市的绿楔。

将商业前广场空间改造为开放、多元的开放空间，可回望远接天际的草坡。中央绿地不设置过重设施及高大乔木，规则绿地和线性旱喷水体强调仪式感，池底雕刻中轴发展的历史变迁。穿越林下休闲阅读空间，环形剧场放置在小型的十字路口。廊架和草坪是公园南侧的主体，公园收尾处结合地形形成坐南朝北的露天剧场，草坪上成列排布的景观灯柱渲染中轴的文化氛围。将科技馆绿地进行改造，打开边缘，引入科技互动的小装置，形成互动性较强的科技公园。

综上，原先被割裂的城市共享空间得到缝合，慢行体系得到优化，城市引入更多绿色空间。

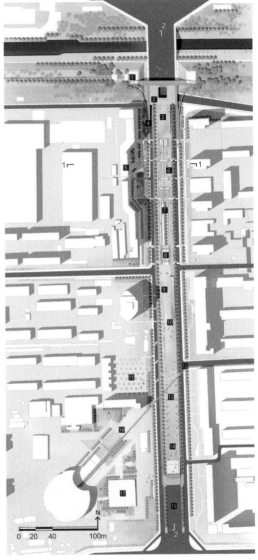

1 元大都城垣遗址公园城墙
2 "城市之窗"高架廊桥
3 天际草坡
4 下行栈桥
5 商业前空间
6 中轴旱喷地雕广场
7 户外休闲空间
8 环形剧场
9 户外草坪空间与露天广场
10 云影停廊
11 生态共享停车场
12 安贞社区公园出入口
13 景观柱阵草坪剧场
14 露天台阶剧场
15 机动车出入口
16 景观柱阵草坪剧场
17 穹顶电影院

中轴设计平面
Design Plan of Axis

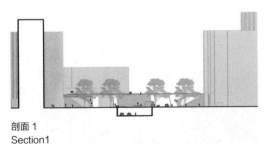

剖面1
Section1

剖面2
Section2

规划设计 PLANNING & DESIGN   133

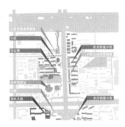

生成过程 1
Design Process 1

生成过程 2
Design Process 2

生成过程 3
Design Process 3

生成过程 4
Design Process 4

商业前广场效果
Perspective of Commercial Plaza

立体交通效果
Perspective of Dimensional Traffic

草地效果
Perspective of Lawn

剧场效果
Perspective of Theater

鸟瞰
Aerial view

5. 乐享生活廊道片区改造

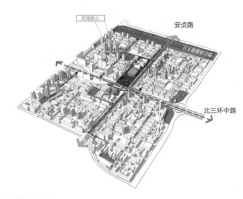

区域位置分析
Regional Location Analysis

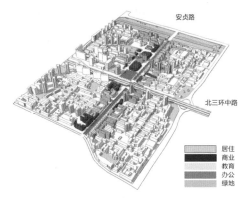

建筑功能类型分析
Building Function Analysis

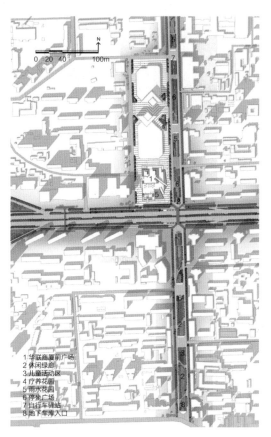

乐享生活片区平面
Design Plan of Living Area

承接上位规划内容，该区域居住用地居多，旨在满足居民日常休闲娱乐、游憩活动的功能需求，形成乐享生活区。

选择区域内华联商厦、木偶剧院等主要的功能性建筑及安贞路、北三环中路等重点交通路段所包围和交叉的部分，同时，考虑现状绿色开放空间的面积及质量，选择区域核心位置作为重点设计地块，具体可划分为街心公园及华联商厦前广场。街心公园是居民日常使用最为频繁的带状绿色开放空间。华联商厦前广场在地块内具有较高的商业及文化价值，激发区域活力。通过分析周边建筑功能类型可知，附近多为居住区建筑，在交通道路的边缘及交叉位置存在一定数量的商业建筑、少量的教育及办公建筑，该区域应综合满足周边不同人群使用需求，故应对每一区域进行具有针对性的功能分析及设计，打造活力四射的街区绿核。

现状主要问题有：①乱停车及占道现象严重；②北侧街心公园大部分绿地空间被地上停车空间取代；③华联商厦前广场被绿篱分隔，没有适宜人们停坐及活动的功能空间；④街心绿地虽质量较高，但已无法满足居民日常功能需求。

针对街心公园，改造策略如下：①街心公园北侧延续至北土城东路，取消内部停车，让出绿地空间，同时将南侧绿地化零为整；②将安贞路中央街心公园下方挖空，作为地下停车空间，可满足现有停车需求。对道路两侧车行时间进行限制，仅在早晚高峰时段允许通行，其余时段为步行及骑行，禁止机动车通行，方便居民日常使用；③将街心花园结合周边使用进行不同功能划分。安贞医院附近设置疗养花园、雨水花园。将华联前广场原本不被人所使用的空间进行下沉处理，安排咖啡休憩等功能，激发空间活力。华联商厦等商业建筑附近设置休息平台，居民集中的部分安排可供儿童活动的场地。

运用以上的设计策略，激活街区绿核，创造崭新的生活方式。

规划设计 PLANNING & DESIGN   135

休闲广场段效果
Perspective of Leisure Square

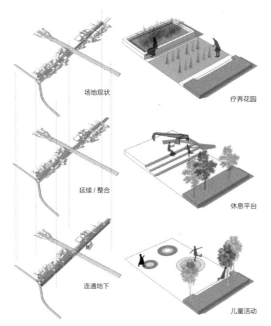

街心花园生成分析
Process of Street Garden Analysis

疗养花园段效果
Perspective of Convalescence Garden

华联商厦前广场效果
Perspective of Hualian Square

华联商厦前广场效果
Perspective of Hualian Square

鸟瞰
Aerial View

6. 邻里绿点设计

此部分对绿色生活共享网络中的邻里绿点进行设计，邻里绿点连接至街区绿核，以实现社区绿色微循环。

设计选取以居住用地为主的地块进行分析，研究邻里绿点选取及设计模式，对花园路街道社区共11个居住小区进行问卷调查。问卷共设置关于社区绿地情况的问题10道，发放问卷共50份。通过对问卷调查结果进行统计，对花园路街道区域内的绿地质量、绿地率、使用情况、可达性、公平性进行评价，在人口密集的区域进行建设，需要有可改造的绿地和未利用的空地。

对现状问题进行剖析，发现场地内缺乏可供老年人及儿童活动的康体活动绿地。因此提出"10min康体生活圈"概念，以体育场地设施为圆点，以其服务半径为轴，以服务半径腹地为圈，打造邻里中心，建立10min体育生活圈。丰富康体生活，激发场地活力。实现出门可达的便捷康体生活，从走街串巷到出门可达，从无处停留到休闲游憩，从无处活动到康体健身。

因此，进行绿点网络构建，将现状可改造绿地、现状空地、人口聚集区绿地、文化属性绿地、现状体育设施潜力绿地叠加，生成区域中的邻里绿点。随后，选取出的邻里绿点。根据居民正常步速步行10min左右，楼下500m左右有健身点、800m左右有健身苑、1000m左右有体育中心进行服务半径核验。按照地块潜力综合得分，对邻里绿点进行评价，按照综合潜力得分进行分类，选择各模式典型地块进行设计。查阅规范及资料，对不同规模的模块进行功能定位，生成总平面，为每个邻里绿点确定主题，并对选取的三个典型地块进行设计。完成各模式典型地块设计，以指导整个规划地块的"10min 康体生活圈"打造。

地块一为健身点，老人之家康体广场，以健身器械、儿童游乐、休闲游憩为主要功能，面积625m^2，分为种植区、活动区、休憩区三个区域，在区块内设置林荫休闲座椅空间。根据年龄设置健身器械，形成出门可达的可满足康体需求的绿地。

地块二为健身苑，七彩生活康体公园，面积约2000m^2，含有球场、体操场，设置阳光草坪，营造社区开放空间，提供球类运动场地，以供社区健身生活。

地块三为体育中心，面积为3hm^2，内设有社区中心，设置管道形儿童活动花园，增加社区公园乐趣，供儿童使用；设置林荫活动场地及球场，激发公园活力。

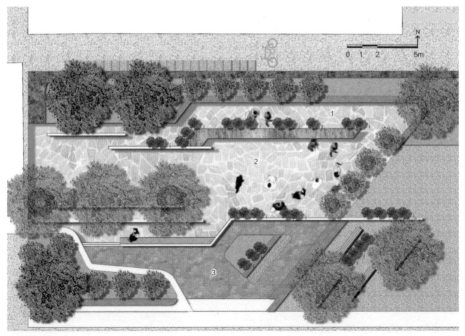

1 休息平台
2 集散广场
3 入口广场

健身点平面
Design Plan of Sport Area

规划设计 PLANNING & DESIGN

喷泉广场效果
Perspective of Fountain

休闲广场效果
Perspective of Square

管道花园效果图
Perspective of Garden

阳光草坪效果
Perspective of Lawn

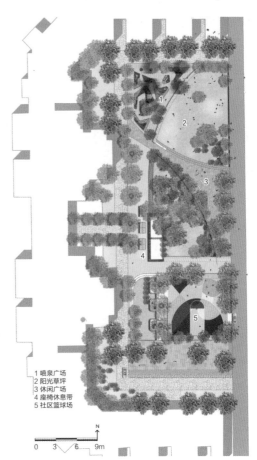

1 喷泉广场
2 阳光草坪
3 休闲广场
4 座椅休息带
5 社区篮球场

健身苑平面
Design Plan of Sport Area

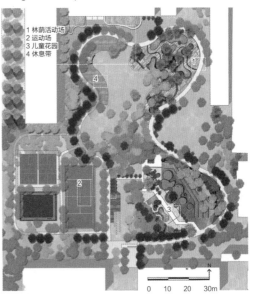

1 林荫活动场
2 运动场
3 儿童花园
4 休息带

体育中心平面
Design Plan of Sport Area

The site is located between North Third Ring Road and the North Fourth Ring Road. The north central axis of Beijing crosses it over. There are the Chinese Ethnic Culture Park, the Beiji Temple Park within the site. It is bordered by the Yuan-Dadu City Wall Relics Park to the south and the Olympic Forest Park to the north. The Asian Games in 1990 and Olympic Games in 2008 had played huge roles in forming the basic urban frame of this area, and which contribute a lot in promoting urban construction of Beijing as a modern international metropolis and a diversified world cultural city with international influences of traditional and modern culture. For the forthcoming winter Olympic Games in 2022, this site will become the focus of Beijing's urban construction again.

There are currently many closed green areas on the site, including the Chinese Ethnic Culture Park and several urban parks. But those parks cannot form a coherent greenway system open to the surrounding communities. Also, scattering the area especially along both sides of the Beijing-Tibet Expressway there are under-developed grey spaces which could be potential spaces to increase the green land and supplements of social services to local communities.

Starting with the conception of "urban double repair ", a green network will be established to collage functional networks, to weave transportation networks, to improve facilities networks, and then to extend cultural networks of the area. Three green belts and a green network core area are planned within the site. As a result, the site will be connected by three vertical and horizontal corridors and the green core area. The new green network will be a loop to connect the sporting areas, the tourist areas and residential areas around the north central axis between the early-built areas in the 1950s and the newly-built areas in the 2000s. Most of the green spaces on this loop will be open to all, aiming to increase the "blood vessels" and realize regional microcirculation.

This plan expects to shape the landscape characteristics of the areas around the north central axis, and also to create an open space system between the ancient city and the modern sporting, tourist and residential areas. Through the construction of the green network system, the urban public space can be optimized to realize the integration of old and new, the inheritance of culture and the sharing of space. The sense of belonging of the site should be improved, the integration of history and future in the open streets will be realized, and a sustainable ecological sharing city is coming soon.

规划设计 PLANNING & DESIGN 139

生活的礼赞
北京北中轴绿网更新

The Praise of Life
The Green Network Renewal of the North Central Axis of Beijing

赵颖、王仲宇、王丽娜、刘丽丽、李膨利、陈思淇
Zhao Ying / Wang Zhongyu / Wang Lina / Liu Lili / Li Pengli / Chen Siqi

地块位于北京城北三环与北四环之间，城市的北中轴从中穿过。基地内有中华民族园、北极寺公园；其南侧紧邻元大都城垣遗址公园，北侧有奥林匹克森林公园。

场地内公园较多，绿量较大，主要有一个专类园中华民族园和多个公园绿地。场地内有良好的生态基底，但缺少社区间绿核，无法形成连贯的绿道系统。随着后奥运时代的大规模、高密度的城市再开发建设以及未来冬季奥运会的举办，该场地将成为北京城市建设的焦点。

方案试图从"城市双修"理念入手，在场地内规划三条贯穿绿带，一块绿网核心区，通过横纵不同的廊道连接，增强场地的生态连通性，最终构建完整的生态绿色网络。

地块位于北三环与北四环间，城市的北中轴从中穿过。基地内有中华民族园、北极寺公园；其南侧紧邻元大都城垣遗址公园，北侧有奥林匹克森林公园，西边与颐和园相呼应。

2008年北京奥运会的举办，对北京城市建设起到了巨大的推动作用，使北京发展成为世界文化名城。随着后奥运时代的大规模、高密度的城市再开发建设以及未来2022年冬季奥运会的举办，该场地将再次成为北京城市建设的焦点。

现状用地性质以居住用地与公共设施用地为主，考虑常住居民的日常生活需求以及外来游客的游览需求。基地内公园较多，绿量较大，有良好的生态基底，但缺少社区间绿核，无法形成连贯的绿道系统。

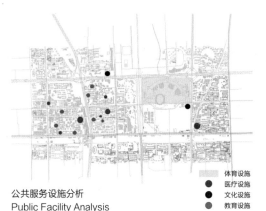

公共服务设施分析
Public Facility Analysis

体育设施
医疗设施
文化设施
教育设施

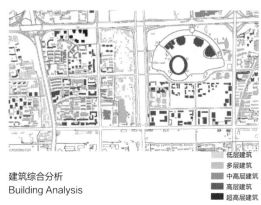

建筑综合分析
Building Analysis

低层建筑
多层建筑
中高层建筑
高层建筑
超高层建筑

公共交通分析
Public transit Analysis

公交站点
地铁站点

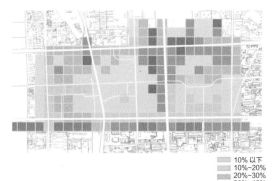

绿化覆盖率分析
Green Coverage Rate Analysis

10% 以下
10%~20%
20%~30%
30%~40%
40%~50%
50%~60%
60%~70%
70%~80%
80%~90%
90% 以上

规划设计 PLANNING & DESIGN 141

因子叠加灰度图
All Factors Gray Scale

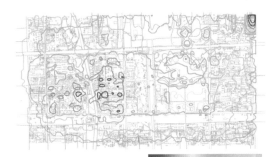

灰度图—去除绿化覆盖率因子
Gray Scale – No Green Coverage

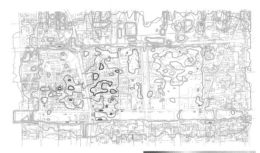

灰度图—去除建筑综合评价因子
Gray Scale – No Building Comprehensive Index

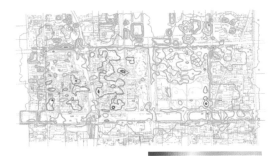

灰度图—去除公共设施因子
Gray Scale – No Public Facility

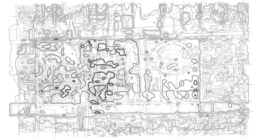

灰度图—去除商业 POI 因子
Gray Scale – No Business POI Index

通过对不同因子的权重分析，发现场地内部京藏高速线性肌理显著，两侧用地再利用潜力大，但目前两侧用地功能单一。此外，设施服务与建筑质量两因子灰度值呈正相关，表明旧街道小区基础服务水平不高；设施服务因子与绿量均布值因子吻合度低，表明绿地的服务能力有提升空间。

从"城市双修"理念入手,修复生态网,拼贴功能网,织补交通网,完善设施网,延展文化网。"城市双修"要点是以改善生态环境质量、补足城市基础设施短板、提高公共服务水平为重点,转变城市发展方式,提升城市治理能力,打造和谐宜居、富有活力、各具特色的现代化城市,让人们在"城市双修"中有更多获得感。

基于此,将规划方向定为修复京藏高速割裂的肌理,提升两侧绿地连通及利用;增加亚运村、奥体南区绿地综合价值;微循环落点到次干道、支路上叠加因子灰度高的区域,激活该地区景观性、生态性;绿地需增加其他因子权重,即增加绿地综合利用度,增加历史地块内的绿地和服务设施,整合现有绿地功能,提升综合服务水平。

在场地内规划三条贯穿绿带,一块绿网核心区,通过横纵不同的廊道连接,增强生态连通性,形成完整的绿色生态网络。三条绿带分别为元大都、京藏高速和中轴旁绿地,一块绿地核心区为奥体中心和奥南商务区。面状公园绿地、点状绿核及线性的街道景观构成了一个完整的绿色网络。通过综合考虑串联旅游景点、沿线公共设施较多、POI指数较高、可实施度高等方面因素,最终确定绿道选线。

策略一
连通的绿色网络,打通区域脉络

策略二
复合的公共空间,提升区域品质

策略三
怡人的步行体系,激发区域活力

设计策略
Design Strategy

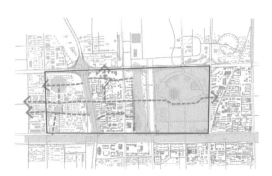

规划结构
Planning Layout

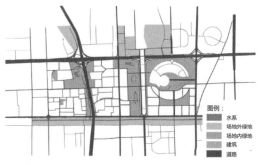

绿地水系规划
Greenland and Water System Planning

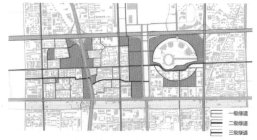

绿道规划
Greenway Planning

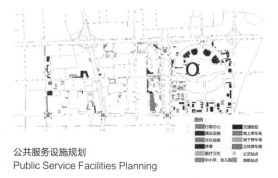

公共服务设施规划
Public Service Facilities Planning

规划设计 PLANNING & DESIGN   143

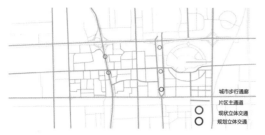

慢行系统—步行规划
Non-motorized Traffic Planning

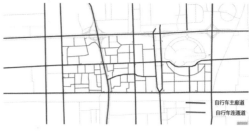

慢行系统—自行车规划
Non-motorized Traffic Planning

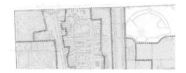

奥林匹克体育中心规划策略
Planning Analysis

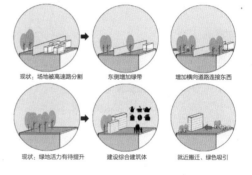

京藏高速路两侧规划策略
Planning Analysis

中轴景观区规划策略
Planning Analysis

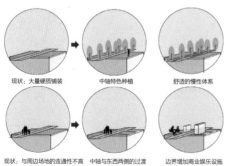

居住小区规划策略
Planning Analysis

1. 北极寺公园及周边场地设计与改造

场地西侧为大片居住区，东侧紧邻京藏高速，京藏高速东侧规划为沿街绿地，场地南侧为元大都城垣遗址公园。

现状交通：场地南接北土城西路，北邻北四环中路，东侧为京藏高速，有地下通道连接南北两侧。场地中部被奥体中路穿过，南侧有地铁健德门站，场地周边有5个公交站。

小月河概况：在1984—1985年，对小月河进行了河道治理，并将河道与西北土城沟组合贯通，小月河从明光寺学院路变为明渠，向北至黄亭子，再向东至祁家豁子，然后一直向北入清河，是清河南部的主要支流。

小月河总长10km，流域面积27km^2，是京城北部唯一南北流向的河流，常水位线与高水位线相差0.5m左右，水深1.5m左右。

小月河虽然经过修整，水体受污染度并不高，但是水体被人工驳岸包围，景观性较差，无法亲水。针对北部和南部特点采用不同方式进行改造。

场地西侧为大片居住区，市场、沿街低矮商铺、宿舍楼、酒店。嘈杂、停车混乱，希望增加商业类建筑，提高场地利用率和景观性，解决场地停车问题。

现状分析
Status Analysis

1-1 断面
Section 1-1

2-2 断面
Section 2-2

1 主入口广场　7 滨水木栈道　13 儿童游乐场
2 滨水广场　　8 花海　　　　14 综合商业街
3 休闲步道　　9 滨水阶梯广场　15 入口广场
4 滨水休闲　　10 休闲草坪
5 绿荫休闲　　11 健身场地
6 林中休闲座椅　12 小型篮球场

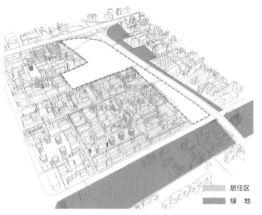

场地外环境分析
Analysis of Site Surrounding

现状交通分析
Analysis of Present Traffic

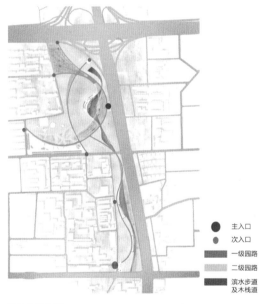

场地交通分析
Analysis of Site Traffic

视线分析
Analysis of View

儿童游乐场效果
Perspective of Playground

滨水休闲段效果
Perspective of Waterfront Leisure Space

2. 京藏高速东侧地块改造设计

设计地块的西侧相隔京藏高速有北极寺公园及小月河,东侧为民族园,南侧为元大都城垣遗址公园,周围不乏大面积的绿地,其内部主要以居住区为主。

现有绿地主要集中在西北侧高速路旁的防护绿地、公园绿地,缺少供附近居民使用的绿地。设计通过增加以绿色空间为主导的网状结构,局部增加面状绿核,完善绿色微循环系统。

京藏高速使得场地与西侧绿地产生割裂,联系不紧密。京藏高速东侧部分改为绿地,容积率较高的建筑予以保留,容积率较低的老旧建筑拆改。现有地下通道和南侧的人行天桥连接京藏高速两侧的

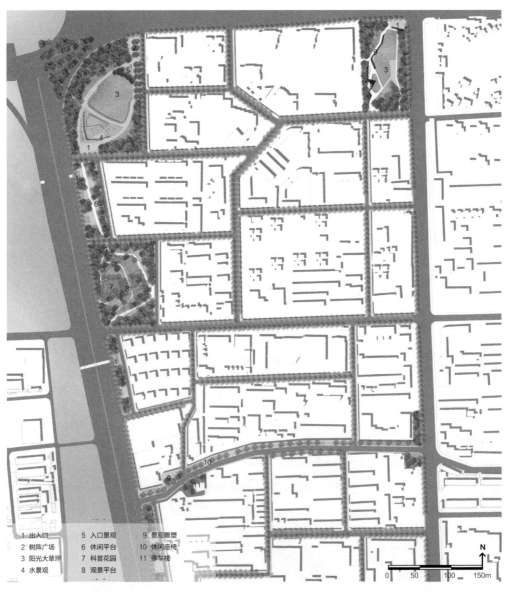

1 出入口　　　5 入口景观　　9 景观雕塑
2 树阵广场　　6 休闲平台　　10 休闲座椅
3 阳光大草坪　7 科普花园　　11 停车楼
4 水景观　　　8 观景平台

京藏高速东侧地块改造平面
Site Plan

地块；内部存在断头路，缺少交通微循环系统。规划希望将路网变密，打通断头路，形成完整道路。同时，还希望将原有商业性较强的小区路通过有效停车策略、拆除局部临时建筑等措施舒缓道路交通压力。

出入口分析
Entrance and Exit Analysis

道路及广场分析
Road and Square Analysis

功能分区分析
Functional Partition Analysis

阳光大草坪效果
Perspective of Large Lawn

科普花园效果
Perspective of Science Garden

街景效果
Perspective of Street View

雕塑公园效果
Perspective of Sculpture Park

3. 中轴

1 主入口广场
2 中心水景
3 中轴广场
4 中心草坪
5 活动广场
6 下沉庭院
7 特色中轴路
8 矩阵树林
9 入口广场
10 旱喷广场
11 缤纷花坛
12 龙形水系
13 自然花海
14 休憩平台

中轴改造平面
Site Plan

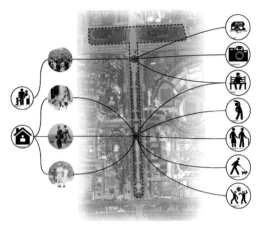

人群分析
Activity Analysis

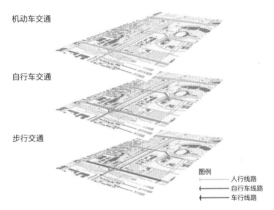

现状交通分析
Current Traffic Analysis

中轴区域的规划目标是规划建设有中轴特色、可识别性高、能满足到访游客与周边居民需求的中轴纪念旅游区域。

此地块位于北京历史中轴线的北段，是承接历史北京与奥运北京的重要段落。区域现状主要存在以下问题：景观尺度过大，功能单一，缺乏基础设计；封闭度高，公众参与度低；相比其他中轴段落，景观性较差、可识别度低；东西连通性低，交通通达性较差；硬质铺装面积过大，生态性较差。结合多次现场调研与资料查阅、案例分析，我们将主要从绿色公共空间和城市景观的公共职能两方面来研究此区域未来发展的潜力与可能性。

第一，要打造可识别系统。创造景观标志物以界定、联系城市中轴，增强场所精神。通过有机地梳理绿地系统框架，使场地具有可识别性，方便到达人群的方向定位，使空间更加有序、高效、舒适。

第二，建立丰富的开放空间及布局层次，区别定义不同的景观空间。提供多样尺度的开放空间系统，保证中轴景观纪念性意义的同时，满足城市公众性活动需求。

第三，提升空间舒适度。增强地块空间的尺度舒适感、提高步行可达性，提高步行、慢行空间品质。采用景观手法有效隔离周边公共交通的噪声干扰，给慢行交通提供有利条件。

第四，生态可持续策略。以景观空间为骨架，综合水处理体系，有效循环利用地表径流，为提高城市整体环境质量提供支持。

规划设计 PLANNING & DESIGN   149

机动车交通
Vehicle Traffic

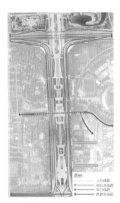

步行交通
Walking Traffic

自行车慢行系统
Cycle Traffic

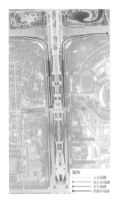

旅游大巴路线
Tour Bus Line

中心活动草坪效果
Perspective of Central Lawn

北段自然休闲区效果
Perspective of Natural Park

活力广场效果
Perspective of Square

中心活动草坪效果
Perspective of Central Lawn

4. 国家奥林匹克体育中心

国家奥林匹克体育中心（以下简称"奥体中心"）始建于 1986 年，于 1990 年第十一届亚洲运动会前建成并正式投入使用。奥体中心建设占地 66hm^2，是一个大型的比赛和训练场地，是我国最主要的体育中心。奥运会后，奥体中心将作为综合性体育中心，以一流的场馆设施和管理服务，承接国内外体育比赛和大型活动。对场地进行一定调研后，我们发现奥体中心内部以专业性较强的体育设施为主，趣味性不足，现状开放度和绿地利用率低。同时，未能与周边新建设施良好衔接。

在对这一亚运主题色彩强烈的地块改造过程中，我们试图对现状室外运动场和停车场进行一定整合，在保证总面积不变的前提下集约出一定空间，利用灵活的空间措施给人们提供多元化的生活体验，使运动、健身、娱乐融合在体育花园之中。

可拆卸式吊床
Removable Hammock

趣味运动设施
Amateur Sports Facilities

帐篷/烧烤
Tent & Barbecue

体育设施
Exercise Facilities

儿童设施
Children Facilities

临时性停车
Occasional Parking

单元模式
Cell Modes

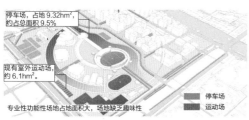

现状问题 1
Existing Issue 1

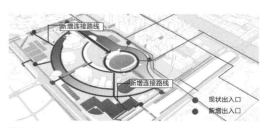

策略 1
Strategy 1

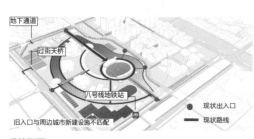

现状问题 2
Existing Issue 2

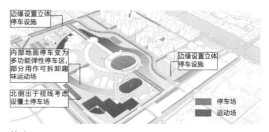

策略 2
Strategy 2

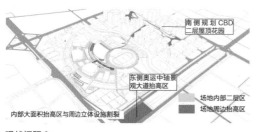

现状问题 3
Existing Issue 3

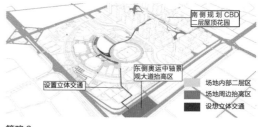

策略 3
Strategy 3

规划设计 PLANNING & DESIGN

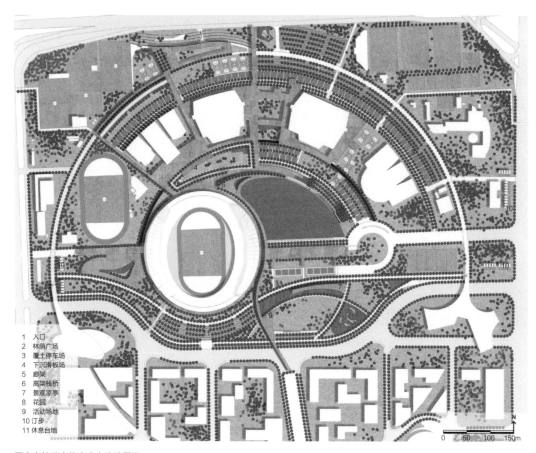

1 入口
2 林荫广场
3 覆土停车场
4 下沉滑板场
5 廊架
6 高架栈桥
7 景观凉亭
8 花园
9 活动场地
10 汀步
11 休息台地

国家奥林匹克体育中心改造平面
Site Plan

覆土停车场效果
Perspective of the Earth-Sheltered Parking Lot

下沉滑板场效果
Perspective of the Sunken Skatepark

景观桥效果
Perspective of the Landscape Bridge

观景平台效果
Perspective of the Viewing Platform

The design site is located between the North 3rd Ring Road and the North 4th Ring Road of Beijing. On the south of the site is the Yuan-Dadu City Wall Relics Park, and on the north of the site is the Olympic Park. It stands in wedge-shaped green land which is close to the shelterbelt and the planned wind tunnel according to the master plan. This location offers great importance for the enhancement of the appearance of the city and the construction of the green ecological network.

The design area of the site covers approximately 4.28 square kilometers. The land is currently used as residential land, green space and public facilities. The problems mainly focus on the following three aspects:

First of all, although the current green space is large, the green space is more homogeneous, the distribution of the green space is not even, and the location of the green space is scattered, which cannot form a coherent green network system. Most of the residential green space is closed, for example, the China Ethnic Culture Park. The Yuan-Dadu City Wall Relics Park and the Olympic sports center green spaces are open, but not including the part of many public services. At the same time, the status of plant has not formed the rich and stable, ecological and landscape plant community. In a word, the existing green space cannot serve the residents well or create good ecological benefits. Secondly, it is the traffic problems. The urban road was planned well, but there are more problems currently. Frequent occupations of road cannot meet the traffic demand and led to traffic congestion. Safety issues such as car accidents are more prominent. Due to the lack of perfect public transport system, serious traffic congestion problem is frequently emerged. The parking lot in the area is unevenly distributed, and the parking space in the residential area occupies a serious public space, which seriously obstructs the traffic. At the same time, a comprehensive slow system has not formed to better facilitate the residents to travel and life. Finally, the cultural aspect. The status quo culture of the site is represented by the Olympic culture, with distinctive features, but its influence on the radiation scope is small, and the regional cultural characteristics are not fully displayed. At the same time, the existing cultural nodes in the site are not connected enough to form a clear cultural tour corridor and enrich the overall cultural atmosphere. In general, the regional advantages of this site is obvious, including abundant current resources, and strong potential plasticity. However, the green spaces are closed and the links between green spaces are insufficient, the transportation is complex, and cultural features are not clear enough. The overall optimization and integration are urged.

According to the superior planning requirements and comprehensive investigations this site is positioned as an Olympic Cultural Center, aiming at reshaping its local culture represented by the Olympic culture, highlighting cultural features, being ecologically oriented and building a green network as a means to update the urban integrated environment to stimulate the vitality of the city. The current green quantity base could be used as a radiation core, some of the roads could be reconstructed and ungraded to form a green corridor to connect scattered small green areas to form a comprehensive green network. Smooth city slow traffic system can ease traffic pressure while creating a comfortable travel environment and then transform existing urban space, increase community public space, and enrich the lives of residents.

Specifically, the North Central Axis connects the two bigger green areas of the Olympic Center and the China Ethnic Culture Park into the green nucleus of the area. And it is appropriately transformed to improve the openness so as to serve the surrounding area better. Through the main roads including North Fourth Ring Road, North Tucheng East Road, and Beijing-Tibet Expressway, this green nucleus of two green areas connects Aonan Business District and several residential areas around it. At the same time, the internal problems in each area will be addressed to transform public spaces, improve the quality of green space, increase spatial connections, and create micro-circulation of urban traffic, create a livable living environment, and improve the quality of life of residents.

In the end, the entire area will form a complete green network system, and will link up with the surrounding green spaces, aiming to improve the structure of Beijing's urban green network, and to achieve sustainable development of the city. The quality of urban life could be improved from various aspects such as transportation, culture, industry and ecology, and as a result the vitality of the development of the area will be promoted. In the future, it is of great significance to improve local ecological environment quality and promote sustainable development of the whole area.

"帧"渗透
北京市中轴绿网更新研究
"Frame" Infiltration
Research of Green Network Renewal in Beijing Central Axis Area

税嘉陵、王倩、张璐韡、王言茗、吴思雨、王芳、宋捷
Shui Jialing / Wang Qian / Zhang Luwei / Wang Yanming / Wu Siyu / Wang Fang / Song Jie

　　设计场地南邻元大都城垣遗址公园，北接奥林匹克森林公园，地处楔形绿地，紧邻防护林带和规划风道，具有重要的文化、生态意义。场地内部以国家奥林匹克中心为主要地块，同时涵盖北京北中轴北端、中华民族园、京藏高速以及多个居民社区，特色鲜明而又情况复杂。交通拥堵、绿化参差不齐等问题严重影响居民的生活品质及区域的生态效益，因此亟待做出统一整改提升。

　　根据上位规划及综合调查研究，此次规划将其整体定位为奥运文化中心区，旨在重塑以奥运文化为代表的区域文化，彰显文化特色，同时以生态设计为导向，因地制宜，充分利用绿量基础，构建功能综合的绿网体系，改善交通现状，打造交通微循环体系，优化整体生活环境。

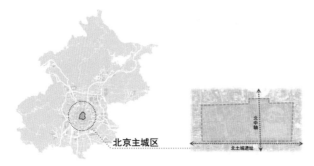

区位分析
Locating

绿地分析
Greenspace Anylsis

绿网覆盖范围
Greenspace Cover

该场地地处楔形绿地，紧邻防护林带和规划风道建立，北有道路绿化廊道。地处奥运文化中心，延续南北方向的城市北中轴，南邻北土城，北接奥林匹克森林公园，西北有规划的风道，东西紧邻居住区。

首先，我们对场地进行了详细分析。对区域绿地分析后，发现规划的绿道系统辐射整个城市，串联不同景观空间，满足多功能的需求。虽然现存的绿网结构覆盖范围较广，但尚未达到网络影响的全覆盖，可以通过增加绿地或强化现有绿地的联系来扩大绿网的影响范围。

对场地用地性质分析后，现状用地性质以居住用地与绿地为主，其次是公共设施用地，所以使用该区域的人群为常住居民与外来游客，因此要考虑常住居民的日常生活需求以及外来游客的游览需求。公共设施用地中体育用地内的绿量较大，可以作为绿核辐射周边；规划区域内的部分道路绿化较好，可以加以改造形成绿网，联系周边现有的绿核或绿廊。规划区域内的绿地分布不均，多集中在体育用地内，社区间绿量少而散，缺少设施完善的小型社区公园；部分道路绿化不足，无法形成连贯的绿道系统。

对场地用地现状绿地分析后，发现现状植物没有形成丰富稳定的植物群落，生态性、景观性较差。对场地文化特色分析后，发现场地现状文化影响辐射范围较小，整个场地缺乏文化特色。

绿地方面：优势是绿地率较大、部分绿地较完整、大面积绿地现状优良；劣势是绿地分布不均、连通性差、部分绿地质量不佳、部分植物群落景观不佳。交通方面：优势是城市干道规划较完善、规划绿道涉及范围较广；劣势是局部机动车道拥堵、慢行舒适度、连通性较差、人车混行现象较严重。文化方面：优势是场地文化特色鲜明；劣势是未充分展现区域文化特色、部分文化场地有封闭性。

最终，对现状分析进行总结，得出了以下结论：①利用绿量基础，完善绿网体系，提升绿网质量，改善生态环境；②改善交通现状，打造微循环体系；③重塑区域文化，彰显文化特色。

接着，分析确定规划设计范围：基于现状绿地、停车场（潜在绿地）、可拆改建筑（潜在）规划出具有生态效应的绿色廊道；基于中轴文化廊道、社区文化廊道、现状绿地，确定出文化游憩的绿色廊道；基于潜在绿地休闲游憩廊道和现状绿地休闲游憩选线，确定出休闲游憩选线；基于潜在绿地上班通勤廊道和现状绿地上班通勤廊道，确定出上班通勤廊道。最后，通过生态廊道、文化廊道、休闲游憩廊道、上班通勤廊道确定出最终的规划设计范围。

根据目标导向与问题导向，提出规划的理念为："帧"循环——CEL，文化（Culture）、生态（Ecology）、生活（Life）。规划策略为修复城市生态，建设绿廊，完善绿地系统，修复利用废弃地，改善场地生态功能；修补城市功能，填补基础设施欠账，增加公共空间，改善出行条件，提升环境品质；塑造绿色空间，完善生态功能，增加雨洪处理系统，创建新型海绵社区；连接绿道网络，通过新增绿廊，连接现状零散、破碎的绿地资源；优化慢行体系，增加社区之间的交流，创造宜居环境；提升文化品位，整合历史、文化资源，在保护的同时提升场地现有文化影响力；激活社区空间，改造老旧小区，增加公共活动场所，建立充满活力的多样化社区空间。

用地性质分析
Land Use Analysis

绿地现状分析
Greenspace situation Analysis

绿地现状分析
Greenspace situation Analysis

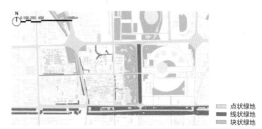

绿地类型分析
Greenspace type Analysis

交通现状分析
Traffic Situation Anylsis

交通现状评价
Traffic Situation Evaluation

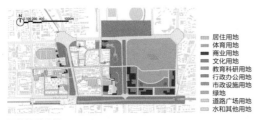

用地性质分析
Land Use Analysis

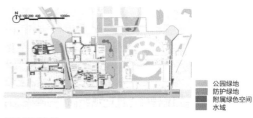

绿地性质分析
Green Space System Analysis

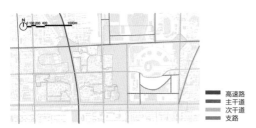

城市机动车道路
Automobile Traffic Analysis

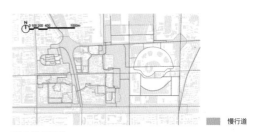

城市慢行系统
Non-motorized Traffic Analysis

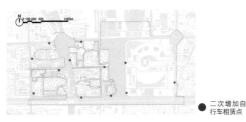

自行车租赁点
Bicycle Rental Point Analysis

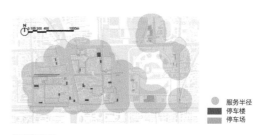

城市停车场
Urban Park Analysis

拆改建筑
Architecture Analysis

商业区调整
Business Zone Planning

文教设施调整规划
Culture And Education Facility Planning

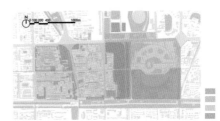

总体分区
Zoning Plan

根据场地的上位规划，确定了场地基本定位——奥运文化中心区。根据场地进行分析总结，提出在生态、生活、文化方面，建立以生态为导向，以构建绿网为手段，更新城市综合环境，激发城市活力的"帧"循环——CEL规划理念。通过修复、塑造、连接、激活的策略实现此目标，修复城市生态、修补城市功能，塑造绿色空间，完善生态功能，连接绿道网络、优化慢行体系，提升文化品位、激活社区空间。

根据现状及定位，将规划范围分为生态绿网区、居住区、商业综合区、奥运综合区。生态绿网区域内，打造完整公园绿廊系统，激活滨水空间、整合破碎绿地、串联绿道斑块、完善绿色网络；居住区区域内，建立海绵社区、改善公共空间质量；商业综合区域内，激活商业街道、建立多样化居民区商业街、构建商业综合体；奥运综合区域内，利用场馆资源、连接周边绿网、提升生态、社会效应。建设绿廊、完善绿地系统，修复利用废弃地，改善场地生态功能，填补基础设施欠账，增加公共空间，改善出行条件，提升环境品质；增加雨洪处理系统，创建新型海绵社区。

规划后，场地绿地更加系统整体、景观层次更加立体丰富。最终构建起集生态、文化、生活于一体的多样综合绿色生态网络。

连接公园体系

营造开敞公共空间

整合街边绿地

打造屋顶花园

打造社区绿地

绿网规划模式
Green Web Planning

打造多功能交通综合体
增加慢行连接

联系区域交通
构筑立体的慢行交通体系

连通被道路阻隔的绿色空间，拆除部分建筑，改造现状停车场

创造慢行公共空间，改造停车场，拆除部分建筑

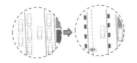
改善慢行道路舒适度；增加植被和基础设施，划定非机动车道

打造主题慢行道路，进行道路绿化
增加沿街立面绿化；增设文化标识
增设体育设施

慢行体系规划模式
Walking System Planning

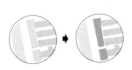
开放民族园商业街；引入特色业态；激活民族园空间

整合沿街商业街；规整立面形式统一广告；开放更多室外空间

多样化居民区商业街；拆除部分临时建筑；整合现状杂乱的商业模式；将商业集中；留出完整空间

新增商业综合体；满足市民物需求；为日常生活提供便利

商业综合区规划模式
Commercial Complex Planning

158 规划设计 PLANNING & DESIGN

规划总平面
Master Plan

规划设计 PLANNING & DESIGN

1. 北极寺小月河区域

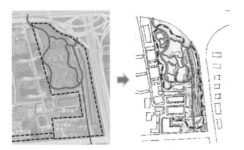

保留现状慢行体系，创造多样慢行体系理

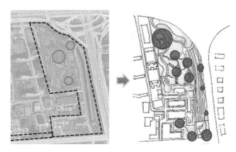

增加活动空间，创造宜人户外活动的场所

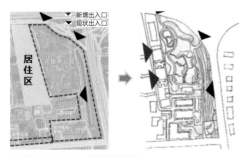

增加朝向居住区入口，增加绿地使用性

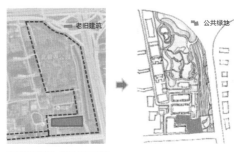

拆除老旧建筑，整合绿地空间

规划策略
Planning Strategy

1 儿童活动场
2 水景（互动性）
3 剧场
4 草阶
5 活动场地
6 老年人活动场
7 入口场地
8 入口
9 开敞广场

北极寺小月河设计平面
Design Plan of Xiaoyue River

规划设计 PLANNING & DESIGN　161

场地包括北极寺公园、小月河、南部的一部分绿地及北极寺干休所，总面积6.6hm²。场地绿量充足，但问题较多，通过重新的规划设计，旨在建立功能健全的城市绿岛，打造城市完整慢行体系中的代表性节点。

通过对场地的现状环境进行分析，得出场地的主要优势和特点是绿量充足、植物长势良好；场地的主要劣势和问题是人车混行严重、缺少宜人的活动场地。

场地主要的设计目标是建立功能健全的城市绿岛，打造整体慢行体系中的代表性节点。设计的总体策略是增强场地连通性，沟通周边区域，构建多样的慢性体系，增加活动场地，提升场地品质。通过整体设计，主要打造了五大场地：儿童活动场地、休闲健身场地、露天剧场区、阳光草阶区及公共开敞区，打造全面的、具有活力的户外开敞区域。

道路规划分析
Road Planning Analysis

高程规划分析
Elevation Planning Analysis

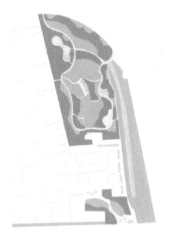

植物空间分析
Plant Space Analysis

儿童活动场
Children Playground Perspective

休闲草阶效果
Leisure Grass Order Perspective

2. 小月河南段区域改造

小月河南段区域位于地块 D 的西南方，右临京藏高速，南接北土城公园，牡丹园和健德门地铁站位于南端两侧，交通便捷。其内部主要为居住用地，生活相对比较便利，文教设施较多，但是缺乏公共休闲活动的空间，同时存在许多破烂的棚户建筑，汽车乱停、乱放现象严重，慢行条件有待提高。现状绿地相对较多，但是分布散乱，不成网络，难以发挥良好的生态效应。本次改造以城市绿网更新为理念，根据场地现状，将潜在及改造的绿地分为点状绿地、线状绿地、块状绿地，结合道路优化，构建形成点、线、面相结合的完整绿网体系，并与周边良好衔接。

成面的块状绿地以小月河南段社区公园为代表，其面积相对较大且规整，是北京规划的五条一级通风廊道的一部分，作为场地生态意义上的大型斑块，是绿网形成的本体，对其的改造策略主要为：拆除原有棚户房，建设为生态公园，增加多种功能，对内服务社区群众，对外作为城市界面，构成良好的城市风貌。

成线的绿色街道以场地内几条主要道路为代表，道路是构成绿网的基本骨架，改造结合开放社区的构建，打通潜在的绿色道路"毛细血管"，连通断头路，完善路网，提高通达性。并且依托现有道路打造舒适的慢行交通体系，形成居住片区绿色交通微循环。

成点的小型公共空间以社区内部某些具有改造潜力的组团绿地为代表，对其的改造通过增加公共设施，丰富空间功能，丰富居民生活，以解决现有公共活动空间不足的问题。

最终经过改造之后，在整个区域内部形成良好的绿网体系，并融入城市绿网。

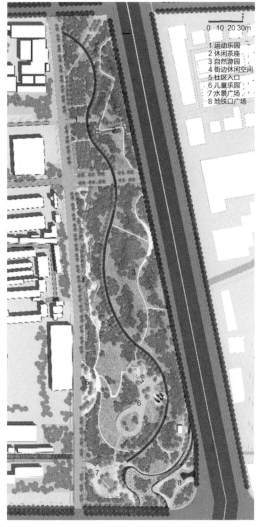

社区公园设计平面
Community Park plan

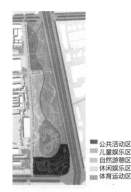

功能分析
Functional analysis

交通分析
Traffic Analysis

景观结构
Landscape structure

竖向设计分析
Vertical design

规划设计 PLANNING & DESIGN

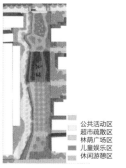

功能分析
Functional analysis

公共活动区
超市疏散区
林荫广场区
儿童娱乐区
休闲游憩区

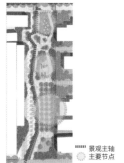

景观结构
Landscape structure

景观主轴
主要节点

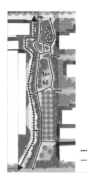

交通分析
Traffic Analysis

主要道路
次要道路

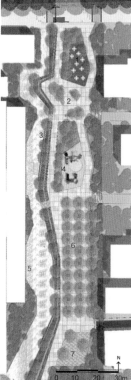

1 棋牌广场
2 休闲空间
3 梯级花带
4 健身空间
5 超市入口
6 林荫广场
7 舞蹈空间

0 10 20 30m

小游园设计平面
Small garden plan

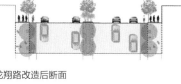

龙翔路改造前断面
Reconstruction section

龙翔路改造后断面
Reconstructed section

小游园局部效果
Part of Small Garden

小游园局部效果
Part of Small Garden

社区公园局部效果
Part of Community Park

社区公园局部效果
Part of Community Park

3. 居住地段改造

根据上位指导"帧"循环——CEL 以生态为主导、文化融入绿色、绿色彰显生活以及此区域是距离奥体中心较近、以居住小区为主的基本认知，得出这一区域的设计目标：将奥运文化和民族文化辐射渗透到周边的居住区，进一步提高居住片区的活力和品质。

这一区域的改造有四点提升策略，分别是渗透提升、共享提升、拆改提升和连通提升。

第一个策略：渗透提升。包括一条文化轴以及奥运文化路和民族文化路的详细改造，提取五环颜色作为分隔柱颜色，形成连贯的自行车网络系统，彰显奥运精神、运动理念和良好的通行环境。

第二个策略：共享提升。把石墙变栅栏，部分围墙开门，使位于小区边缘的绿色空间实现居民共享。

第三个策略：拆改提升。包括社区公园1、社区公园2、社区绿色空间、社区商业步行广场和社区街旁绿地。

社区公园1：设置连贯的自行车道和绿化将社区公园与西部的小月河公园相连，并塑造以波浪元素为主题元素的运动型社区公园景观。

社区公园2：通过介入旋转元素和各色彩的廊架灯柱元素，形成五色主题的休闲社区公园，色彩提取自奥运五环。

社区商业步行广场：建设小型沿街广场空间，创造四季花园主题。

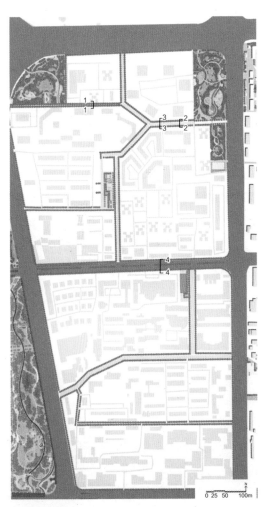

居住地段设计平面
Design Plan of Residential District

1-1 剖面
Section 1-1

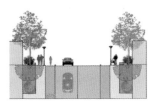

2-2 剖面
Section 2-2

3-3 剖面
Section 3-3

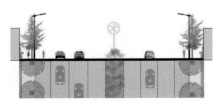

4-4 剖面
Section 4-4

规划设计 PLANNING & DESIGN

社区街旁绿地：建设停车楼，与社区公园 1 和社区公园 2 同样有着疏解路边停车的功能。

第四个策略：连通提升。通过打开道路中央围栏并设置引导路标的方式，连通社区绿色空间与民族园，连通社区绿色空间与元大都城垣遗址公园，连通小月河公园与社区公园。

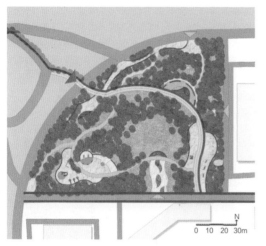

社区公园设计平面
Design Plan Of Community Park

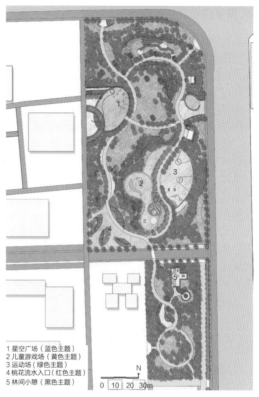

1 星空广场（蓝色主题）
2 儿童游戏场（黄色主题）
3 运动场（绿色主题）
4 桃花流水入口（红色主题）
5 林间小憩（黑色主题）

五色社区公园设计平面
Design Plan Of Five Color Community Park

蓝色主题区：星空广场效果
Blue Theme Area

黄色主题区：儿童游戏场效果
Yellow Theme Area

红色主题区：桃花流水入口场地效果
Red Theme Area

社区公园鸟瞰效果
Community Park Aerial View

4. 奥体中心的轴线改造

奥体中心轴线部分是北京南北轴线的组成部分，是实体的北中轴的末端，地处城市规划的奥运文化中心区内。结合规划定位以及场地周边地块的功能布局，认为整个区域具有丰富的旅游资源，可以彰显我国的奥运文化、民族文化、历史文化，是一个可以充分体现国家形象、城市形象的区域。

整个改造区域主要的使用人群为游客，其次是周边居民和极限运动的青少年，其中游客占据了这个区域大多数的使用时间。轴线周边的机动车交通现状良好，慢行交通以南北方向为主，仅在必要处设置东西向慢行交通。

结合规划定位以及现状分析，将改造区域定位为"承南启北、连东接西"——南承城市中轴、历史文化，北启奥运文化、奥运旅游资源，东连民族文化、居住用地，西接奥运文化和体育、商业用地。

结合规划定位，提出改造目标：重塑场地功能——整合所需功能、合理配置；激发场地活力——提升场地参与度、趣味性；串联场地交通——重新定义轴线边界、打造舒适慢行。旨在将整个场地与周边区域连通，打造更大范围内的绿核。

鉴于场地的现状建设程度较大，且功能相对完善，提出了以下几条改造策略：①轴线周边机动车交通下穿——提升整个区域与周边地块的慢行交通的联系；②改善轴线与周边城市用地的肌理关系——明确城市轴线，并将轴线与周边环境很好地融合在一起；③提升轴线的景观与功能——在满足对应使用人群应有的功能之外，提升空间的景观性，进一步彰显其重要的城市地位与形象。

现状轴线西立面图
Current west elevation

现状轴线东立面图
Current east elevation

机动车下穿入口效果图
Perspective of Stereo-Traffic

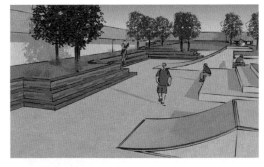

滨水生态段效果图
Perspective of Waterfront

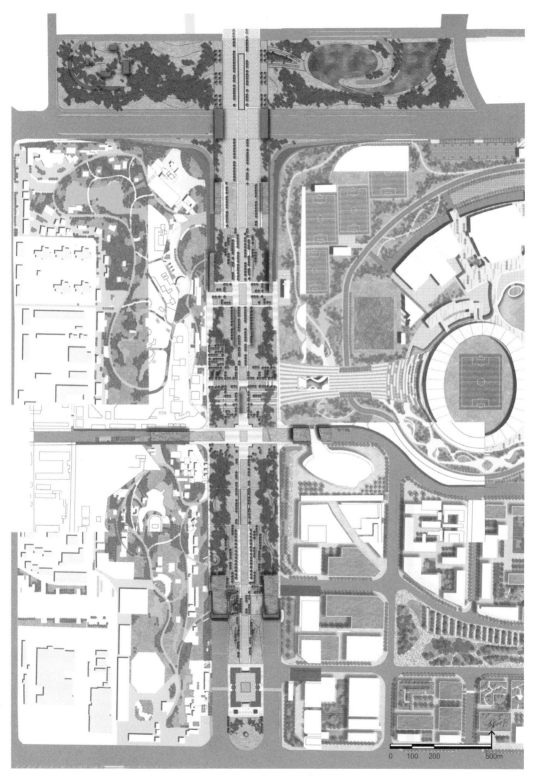

奥体中心的轴线设计平面图
Design Plan of Olympic Central Avenue

5. 奥体中心改造

奥体中心位于此次设计场地中的文化集中区，此次设计将通过场馆功能升级和户外环境改造，建设成为"体育氛围浓厚、赛事举办一流、群众体育活跃、绿化空间宜人"的公共体育活动集聚区，主要承担承办国内外顶级体育赛事、满足市民健身休闲要求、开展青少年业余训练和引领体育产业发展四项功能。

交通组织
Traffic Organization

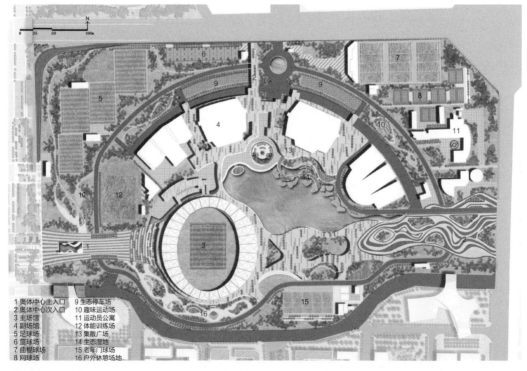

奥体中心设计平面
Design Plan of the Olympic Center

规划设计 PLANNING & DESIGN   169

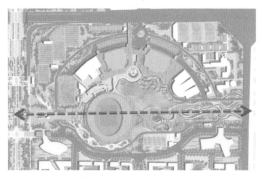

场馆改造分析
Stadium Reconstruction Analysis

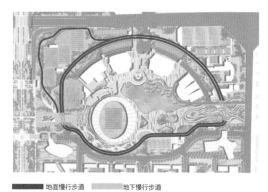

慢行体系分析
Slow System Analysis

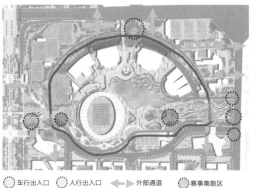

交通组织分析
Traffic Analysis

运动场地分析
Sports Ground Analysis

趣味运动场地立面
Fun Sports Ground Elevation

儿童场地效果
Children's Site Perspective

湿地区域效果
Wedand District Perspective

02 研究团队

RESEARCH TEAM

核心研究团队

Core Researchers

王向荣

1963年生，甘肃人，博士，北京林业大学园林学院副院长、教授、博士生导师，研究方向为风景园林规划与设计。

林箐

1971年生，浙江人，博士，北京林业大学园林学院教授、博士生导师，北京多义景观规划设计研究中心主任，法国凡尔赛国立景观学院访问学者，研究方向为风景园林规划设计理论与实践。

刘志成

1964年生，江苏人，博士，北京林业大学园林学院教授、园林设计教研室主任，研究方向为风景园林规划与设计理论。

李倞

1984年生，河北人，北京林业大学园林学院副教授，研究方向为现代风景园林设计理论与实践、景观基础设施。

王沛永

1972年生，河北人，博士，北京林业大学园林学院副教授，风景园林工程教研室，研究方向为风景园林工程与设计，海绵城市建设及城市绿地用水的可持续设计研究。

钱云

1979年生，江苏人，博士，北京林业大学园林学院城市规划系副教授，研究方向为城市风景环境规划设计，住房与社区研究。

段威

1984年生，武汉人，北京林业大学园林学院副教授，北林园林规划设计研究院风景建筑研究中心副主任，清华大学建筑学院博士。

张云路

1986年生，重庆人，博士，北京林业大学园林学院园林设计教研室副教授，研究方向为风景园林规划设计。

尹豪

1976年生，山东人，博士，北京林业大学园林学院副教授，研究方向为现代园林设计理论，植物景观营造、生态规划与设计。

- 倪锋——北京市规划委员会西城分局　局长
  Ni Feng- Director of Xicheng Sub-bureau of Beijing Planning Commission

- 李战修——北京创新景观园林设计有限责任公司　总经理
  Li Zhanxiu – General Manager of Beijing Innovative Landscape Design Co., Ltd.

- 朱育帆——清华大学建筑学院景观系　副系主任
  Zhu Yufan -Vice Dean of Landscape Department, School of Architecture, Tsinghua University

- 王超——北京京林园林绿化工程有限公司　董事长
  Wang Chao - Chairman of Beijing Jinglin Landscaping Engineering Co. LTD

- 周浩——北京京林联合景观规划设计院　院长
  Zhou Hao - President of Beijing Jinglin Joint Landscape Planning and Design Institute

特邀专家
Invited Experts

赵颖、王仲宇、王丽娜、刘丽丽、李膨利、陈思淇
Zhao Ying/Wang Zhongyu/Wang Lina/Liu Lili/Li Pengli/Chen Siqi

曾筱雁、冯君明、李承玺、刘亚男、孟城玉、王宏达、王科
Zeng Xiaoyan/Feng Junming/Li Chengxi/Liu Yanan/Meng Chengyu/Wang Hongda/Wang Ke

税嘉陵、王倩、张璐韡、王言茗、吴思雨、王芳、宋捷
Shui Jialing/Wang Qian/Zhang Luwei/Wang Yanming/Wu Yusi/Wang Fang/Song Jie

皇甫苏婧、石渠、王美琳、王宇泓、于雪晶、张真瑞、赵人镜
Huangfu sujing/Shi Qu/Wang Meilin/Wang Yuhong/Yu xuejing/Zhang Zhenrui/Zhao Renjing

贾子玉、王念、巴雅尔、周超、孙悦昕、卓荻雅、霍曼菲
Jia Ziyu/Wang Nian/Ba Yaer/Zhou Chao/Sun Yuexin/Zhuo Diya/Huo Manfei

研究生团队
Postgraduates

王思杰、蒋鑫、宋怡、邢露露、姜雪琳、韩炜杰、张希
Wang Sijie/Jiang Xin/Song Yi/Xing Lulu/Jiang Xuelin/Han Weijie/Zhang Xi

许少聪、杨宇翀、罗雨薇、逯羽欣、刘心梦、耿菲
Xu Shaocong/Yang Yuchong/Luo Yuwei/Ti Yuxin/Liu Xinmeng/Geng Fei

刘涵、高敏、阎姝伊、刘喆、胡凯富、刘峥
Liu Han/Gao Min/Yan Shuyi/Liu Zhe/Hu Kaifu/Liu Zheng

本书编辑工作

排版校对　许少聪　杨宇翀　逯羽欣　罗雨薇　刘心梦